BISPHENOL A EXPOSURE AND HEALTH RISKS

Edited by **Pinar Erkekoglu**
and **Belma Kocer-Gumusel**

For EU product safety concerns:
IN TECH d.o.o., Prolaz Marije Krucifikse Kozulić 3, 51000 Rijeka, Croatia,
info@intechopen.com or visit our website at intechopen.com.

Meet the editors

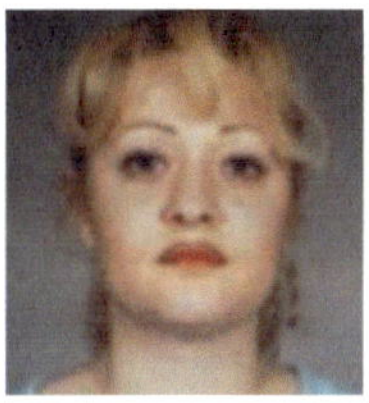

Pınar Erkekoglu graduated from Hacettepe University, Faculty of Pharmacy, and received her MSci and PhD degrees in Toxicology. She worked in Université Joseph Fourier and CEA/INAC/LAN (Grenoble, France) during her PhD thesis. She worked as a postdoc and a visiting associate in MIT Biological Engineering Department. She is currently working as an associate professor in Hacettepe University, Faculty of Pharmacy, Toxicology Department. Her main interests are endocrine-disrupting chemicals and reproductive disorders. She has published more than 120 papers in national and international journals. She is a European Registered Toxicologist.

Belma Kocer Gumusel graduated from Hacettepe University, Faculty of Pharmacy. After receiving her PhD degree, she worked as a postdoctoral fellow at Joseph Fourier University and in CEA/INAC/LAN. She is currently working as a full professor and serves as the department head in Hacettepe University, Faculty of Pharmacy, Department of Pharmaceutical Toxicology. Her main interests are the reproductive disorders and endocrine diseases. She has published more than 140 papers in national and international journals. She is a European Registered Toxicologist. She is a member of many international and national scientific societies.

Contents

Preface

Bisphenol A (BPA) is a synthetic compound, which is employed to make certain plastics (i.e., polycarbonates) and epoxy resins. BPA hardens and clears the polycarbonate plastics and is present in a variety of common consumer goods, such as water bottles, sports equipment, and thermal papers such as that used in sales receipts and food and beverage cans.

BPA is suggested to have endocrine-disrupting properties. It is mainly classified as an estrogen-like endocrine-disrupting chemical, and it interacts with estrogen receptors. In the last decade, attention has arisen in scientific communities that it is not safe to use this chemical in mainly polycarbonate plastics, due to its wide variety of toxic effects. In January 2011, the use of BPA in baby bottles was forbidden in all European Union countries. In July 2012, the US Food and Drug Administration (FDA) banned BPA from baby bottles and sippy cups. By the end of 2014, 12 US states banned BPA from children's bottles and feeding containers.

BPA is estimated to be present in the biological fluids of 93% of humans in detectable amounts. Exposure to BPA starts in prenatal period, which is the critical period for its toxic effects on different organ systems. BPA can affect mainly reproductive system in both males and females, particularly when the organism is exposed to this chemical in early years of life. Moreover, it is suggested to cause toxic effects in the liver, kidney, brain, and thyroid. BPA is also associated with breast cancer development in rodents. Exposure to BPA and subsequent development of breast cancer in humans is unclear. BPA can lead to toxicity even when taken at very small amounts, and "low-dose toxicity" is mainly considered to be critical.

Due to its toxic effects, BPA is suggested to be replaced by different bisphenols, like bisphenol S and bisphenol F, in "BPA-free products." However, these compounds are also shown to exert toxicity, and they are not good alternatives for BPA. Therefore, the industry continues to produce and use BPA in high quantities.

In *Bisphenol A Exposure and Health Risks*, we will mainly focus on the toxic effects of BPA. Throughout this book, the readers will obtain information on the effects of BPA on different systems, particularly on both male and female reproductive systems. They will also get information on the prenatal and postnatal effects of this endocrine disruptor. We believe that readers will get qualified scientific knowledge and a general overview of the toxic effects of BPA exposure and its consequences from this book.

Assoc. Prof. Dr. Pinar Erkekoglu and Prof. Dr. Belma Kocer-Gumusel
Hacettepe University, Faculty of Pharmacy
Department of Toxicology
Ankara, Turkey

Section 1

Effects of Bisphenol A on Different Systems

Chapter 1

Bisphenol A: Understanding Its Health Effects from the Studies Performed on Model Organisms

Papiya Ghosh, Sohini Singha Roy,
Morium Begum and Sujay Ghosh

Additional information is available at the end of the chapter

http://dx.doi.org/10.5772/intechopen.68971

Abstract

Bisphenol A [4,4′-(propane-2,2-diyl)diphenol] (abbreviated as BPA) is a synthetic xenoestrogenic chemical and endocrine disruptor. It is a most common plasticizer that is used widely to produce epoxy resin and polycarbonate plastics, enters the living system through food and water contamination and generates health hazards. Researches are being conducted to explore the adversity that BPA exerts in living body, and for this reason, model organisms are of scientific choice. Rodents, zebrafish, *Drosophila*, nematodes, crustaceans and echinoderms are being used for monitoring the effect of BPA on their life history traits, nervous system, endocrine system, reproductive systems, behaviour, etc., which could help us to anticipate what kind of challenges BPA is putting in human life. This systematic review is focused on the latest research trend on BPA toxicity on different model organisms.

Keywords: bisphenol A, rodent, zebrafish, *Drosophila melanogaster*, invertebrates, reproductive system, life history traits, developmental defects, gene expression

1. Introduction

Bisphenol A [4,4′-(propane-2,2-diyl)diphenol] (abbreviated as BPA) is a synthetic xenoestrogenic chemical and endocrine disruptor [1–3] widely used in dentistry, food packaging and as lacquers to coat food cans, bottle-tops and water pipes since the 1960s. It is a most common plasticizer that is used widely to produce epoxy resin and polycarbonate plastics. It was first synthesized by Dianin in 1891 and was investigated for potential commercial use in the 1930s during a search for synthetic estrogens. BPA enters the living system inconspicuously through various routes, particularly through food and water contamination, and creates multitude of imperilments at cellular, molecular and genetic level. The EC_{50} and LC_{50} values of BPA range from 1.0 to 10 mg/L (Environment Canada 2008), and BPA is declared as 'moderately toxic' and 'toxic' to aquatic biota by the European Commission and the United States Environmental Protection Agency (US EPA), respectively [4], Commission of the European Communities 1996]. Moreover, environmentally relevant concentrations (12 mg/L or lower) of BPA were also found to be harmful as far as wildlife is concerned [5]. BPA exerts its effect through direct binding to estrogen receptor (ER) in a wide range of species that includes invertebrates, fish, amphibians, reptiles, birds and mammals [6]. BPA binds both ERα and ERβ receptors, with approximately 10-fold higher affinity to ERβ [7].

The toxicokinetics of BPA exposure reveal that after oral administration in human, BPA is metabolized rapidly in the intestine and liver. BPA is not completely metabolized via Phase I reactions, but it is rapidly conjugated with glucuronic acid (Phase II metabolism) to produce non-active BPA-glucuronide in the gut wall and liver. Little amount of BPA also reacts with sulphate to form BPA-sulphate compound. The formation of BPA conjugates with other chemical moieties is a detoxification process [8, 9]. The BPA conjugates formed in the liver reach the kidney through blood circulation and then excreted in the urine with terminal half-lives of less than 6 hours [10, 11]. According to a declaration made in 2010 by U.S. Food and Drug Administration, exposure to BPA is alarming because of possible health hazards it exerts on brain, behaviour and prostate gland of foetuses, infants and children. The European Food Safety Authority (EFSA) reviewed new scientific information on BPA in the years 2008, 2009, 2010, 2011 and 2015, concluding on each occasion the known level of exposure to BPA to be hazardous. In February 2016, France announced that it intends to propose BPA as a REACH Regulation candidate substance of very high concern (SVHC).

Owing to difficulty in doing research on human subjects, researchers prefer to use model organisms to test the toxic effect of xenobiotic agents in living system. This approach is also popular in the research on BPA as the agent is ubiquitously present in our 'plastic wrapped world' and no perfect control subject could be obtained in natural environment. Several model organisms from different taxa are in use for studying the effects of BPA on their life history, morphological traits, reproductive functioning, neural functioning and behaviour. The outcome of these studies helps to anticipate the probable adversity that BPA inflicts in human body. Keeping all these factors in mind, a critical review on latest research works is presented here to understand the deleterious effects of BPA exposure on different vertebrate and invertebrate model organisms that could facilitate the understanding of human health hazards due to exposure to this xenoestrogen and endocrine disruptor BPA.

2. Vertebrate model animals

2.1. Studies on rodents

Owing to close taxonomic proximity, rodents, including rat, mice and hamster, resemble most with of humans among all other commonly used vertebrate models, and many symptoms of human conditions can be replicated in mice and rats. For that reason, rodents occupy the most preferred model animal in biomedical research, and keeping pace with this global trend, BPA researchers also rely on rodents to unravel the BPA effects on mammals.

2.1.1. *Effects on reproductive system*

Almost all xenobiotic agents have been tested for their toxicity in rodents to anticipate the probable effects on human body owing to taxonomic closeness of rodents and human as primate. There is extensive evidence that BPA imperils development of reproductive system in male rats and mice, although there appear to be species, strain and dose differences in the sensitivity of specific outcomes to BPA [3]. There are numerous studies of the effects of low doses of BPA on the development of the female and male reproductive organs in rats and mice. Findings include chromosomal abnormalities in oocytes in females [12, 13] and long-term effects on accessory reproductive organs that are not observed until mid-life, such as uterine fibroids and para-ovarian cysts [14]. In Newbold's study [14], outbred female CD-1 mice were treated on days 1–5 with subcutaneous injections of BPA (10, 100 or 1000 μg/kg/day). At 18 months of age, ovaries and reproductive tract tissues exhibited significant increase in cystic ovaries and cystic endometrial hyperplasia in the BPA-treated group. Progressive proliferative lesion of the oviduct and cystic mesonephric (Wolffian) duct remnants was also seen in BPA-treated groups [15].

The effect of BPA on male reproductive organs and function includes decrease in testosterone secretion [16] and sperm production [17, 18]. Impacts on other reproductive structures include reduction in the size of the epididymis at a dose of 2 ng/g and enlargement of the size of prostate ducts in the male foetuses when pregnant females were exposed to a dose of 10 μg/kg BPA/day [19, 20]. These findings are consistent with effects of low doses of positive control chemicals, such as diethylstilbestrol (DES) and ethinyl estradiol. Moreover, the testicular function impairment includes germ cell sloughing, disruption of the blood-testis-barrier and germ cell apoptosis [21, 22].

Impairment in testicular function is also evident in other studies [23, 24, 25]. The effects of BPA resemble more or less the estrogenic effects on the testes [18, 26, 27] with reduction in daily sperm production [28], deformed acrosomal vesicles, acrosomal caps, acrosomes and nuclei of the spermatids. Tohei et al. [29] reported that plasma concentration of testosterone was decreased, and LH was increased in rats after administration of BPA. Testicular content of inhibin was decreased. The testicular response to human chorionic gonadotropin (hCG) for progesterone and testosterone release was also decreased in BPA-treated rats. These results suggest that BPA directly inhibits testicular functions by disrupting the pathway of negative feedback regulation.

Studies have revealed that BPA exposure also affects the female systems, and it is found to be associated with a number of anomalies like polycystic ovarian syndrome [30], endometriosis [31] and anovulation. Studies have also been conducted to evaluate effects of BPA on development of mammary gland. *In utero* exposure to 25 and 250 μg BPA/kg body weight showed changes in the mammary glands of CD1 mice, including a significant increase in the percentage of gland ducts, terminal ducts, terminal end buds and alveolar buds at 6 months of age [32]. Perinatal exposure to 25 and 250 ng BPA/kg body weight showed increased area of terminal end buds relative to the gland ductal area [33]. Studies in both rats and mice have shown that BPA induces change in mammary gland morphology that may predispose animals to develop cancer [34, 35]. **Table 1** shows the summary of results from experiments on reproductive system of laboratory rodents.

Affected area	Model; time and route of exposure	Effect	Citation
Ovary	Mice; developmental, pellet implantation	Disruption of early oogenesis	Susiarjo et al. [12]
Ovaries and reproductive tract tissues	CD-1 mice; developmental, injection	Increase in cystic ovaries and cystic endometrial hyperplasia	Newbold et al. [14]
Mammary gland	Mice, Rats; developmental, injection, minipump	Enhanced growth and differentiation	Markey et al. [32]; Munoz-de-Toro et al. [33]; Soto et al. [13]; Durando et al. [34]; Murray et al. [35]
Testes	Mice, rats ; developmental, adult, oral, injection	Decreased testosterone secretion and sperm production ; deformed sperm with reduced motility	Akingbemi et al. [16]; Aikawa et al. [17]; Toyama et al. [18]; Al-Hiyasat et al. [23]; Chitra et al. [24]; Sakaue et al. [26]
Seminiferous tubules	C57BL/6 mice; adult, oral	Disrupted	Takao et al. [25]
Blood	Rats, adult, oral	↓ Plasma testosterone and ↑ LH	Tohei et al. [29]
Prostate gland	CF-1 mice, CD-1 mice; developmental, oral	↑ weight, ↑ prostate duct volume	Thayer et al. [27]; Timms et al. [20]
Epididymis	CF-1 mice; developmental, oral	↓ size	vom Saal et al. [19]

Table 1. Summary table of the various effects of BPA exposure on reproductive system of laboratory rodents.

2.1.2. Effects on nervous system

BPA has both indirect and direct effects on the nervous system. Since gonadal hormones in conjunction with other neurotrophins regulate cell death, neuronal migration, neurogenesis and neurotransmitter plasticity [36], BPA, in disrupting sex hormone functions, can affect brain development. Estrogen plays a major role in development and differentiation of certain parts of male and female brains. Male and female brains are exposed to different amounts of estrogen during development, and this appears to shape some regions of the brain differently. One of these regions is the hypothalamus, which controls a variety of basic functions including hunger, mood and sex drive. Due to its estrogenic and antiandrogenic activities, BPA can interfere with the dimorphic development of the neuronal networks of male and female brain regulating [37] the activation of hypothalamic estrogen or androgen receptors, testosterone-activating enzymes and hippocampal aromatase expression [38].

As BPA disrupts thyroid function, it can also affect the development of the nervous system because thyroid hormones regulate prenatal and neonatal development of the brain [39]. Juvenile hypothyroidism due to BPA exposure leads to diminutive dentritic growth in hippocampal neurons of rat brain, resulting in cognitive defects including impaired memory, defective perception and attention problems [40]. In a prenatal study [41] of brain development in mice treated with BPA in a dose 20 μg/kg, body revealed decrease in growth in the ventricular zone of the BPA-treated offspring, whereas in the cortical plate, growth was increased. In addition, the expression of thyroid Receptor gene TRα (and other genes) was significantly upregulated in the cortical area of the BPA-treated group. BPA induces cortical plate growth via upregulation of the thyroid pathway. In doing so, BPA might have disrupted normal neocortical development by accelerating neuronal differentiation and migration. BPA exposure may also interfere with the development and expression of normal sex differences in cognitive function, via inhibition of estrogen-dependent hippocampal synapse formation in female rat [42] and testosterone-induced hippocampal synapse formation in male mice [43].

In addition, BPA may directly cause neurodegeneration. BPA enhances hydroxyl radical formation in the rat brain [44], and it is induced by 1-methyl-4-phenylpyridinium ion (MPP+) [45]. This leads to neurodegeneration of the *substantia nigra* and produces acute Parkinsons like symptoms. In this study, 10 μM BPA was infused into the rat striatum to generate OH radical, and *in vivo* micro-dialysis technique was used for evaluating toxic effects on nervous tissues. In another study [46], BPA was shown to increase intracellular reactive oxygen species at a concentration of 1, 10, 25 and 50 μmol/L and induce apoptosis at a concentration of 100 μmol/L in mesencephalic neuronal cell culture. Besides, BPA has a significant impact on the dopaminergic system and hippocampal-associated cognitive functions. **Table 2** represents the various observations on the nervous system of laboratory rodents exposed to BPA.

2.1.3. *Effects on chromosomes*

Recently, researches have unravelled the fact that maternal exposure to a very low dose (20 ng/g body weight) of BPA disrupts alignment of chromosomes during meiosis in the embryonic oocyte during formation of the primary follicles. This abnormality was also observed in mice that were housed in polycarbonate cages and that were provided water in polycarbonate bottles that had been damaged by exposure to a harsh detergent during washing [47]. This finding suggests that exposure to BPA during the time that meiosis resumes in the mid-cycle surge by luteinizing hormone (LH) can result in an increase in foetal aneuploidy and subsequent spontaneous abortion in humans [47]. The effect of BPA on aneuploidy has also been examined in cell culture [48–51]. In the study by Tsutsui et al. [48, 49], treatment of Syrian hamster

Affected area	Model; time and route of exposure	Effect	Citation
Brain	Mice; developmental, injection	↓ growth of ventricular zone, ↑ cortical plate growth	Nakamura et al. [41]
Hypothalamus	Mice, rats; developmental, injection	Affect sex differences in brain development	Negri-Cesi [38]
Hippocampus	Sprague-Dawley rats; adult, injection	Inhibits synapse formation at CA1 area	MacLusky et al. [42]; Leranth et al. [43]
Striatum	Rat; adult, infusion	Neurodegeneration of *substantia nigra*	Obata and Kubota [44]

Table 2. Summary table of the various effects of BPA exposure on nervous system of laboratory rodents.

embryo cells with BPA (100 μM) for 48 hours resulted in statistically significant increases in the percentage of aneuploid metaphases with chromosome losses. Reports are also available that revealed delay in the meiotic cell cycle, possibly by a mechanism that degrades centrosomal proteins and thus perturbs the spindle microtubule organization and chromosome segregation in mouse oocyte during meiosis. When cultured cells were exposed to BPA during the transition from meiosis-I to meiosis-II, a delay in meiosis-I had been observed. This transition phase usually lasts for 8–10 hours in mice, but for BPA-exposed culture, 53% of cells remained in meiosis-I. Insignificant counts of cells were found in anaphase [52].

2.1.4. Effects on behaviour

With inevitable effects of BPA on nervous system, behavioural patterns of rodents are reported to be affected by BPA exposure. An increase in defensive aggression was reported in the offspring of male Sprague-Dawley rat whose mother was offered oral BPA dose (40 μg/kg/day) throughout gestation [53]. In addition, increased aggressiveness (using a composite score of aggression) in male CD-1 mouse offspring was evident as a result of oral administration of low dose of BPA (2 and 20 ng/g of body weight) to pregnant females on gestation days 11–17 [54, 55].

A series of studies demonstrated that prenatal and neonatal exposure to BPA upregulates activities of the dopamine system and induced hyperactivity among the experimental rat [56]. Support to this primary report came from the study [57] that revealed prenatal and neonatal exposure of mice to BPA caused upregulation of dopamine D1 receptors, produced hyperlocomotion and increased rewarding responses induced by methamphetamine. Narita et al. [58] demonstrated that exposure of mice to BPA during either organogenesis or lactation, but not implantation and parturition, significantly enhanced the morphine-induced hyperactivity and rewarding effects. In a rat model, Ishido et al. [59] demonstrated that neonatal exposure to BPA (87 nmol/10 μl/rat) caused significant hyperactivity at 4–5 weeks of age, and significantly decreased gene expression of dopamine transporter at 8 weeks.

Negishi et al. [60] demonstrated that BPA impaired both passive and active avoidance learning among offspring of Fisher 344 rats that were fed a low dose of BPA (0.1 mg/kg/day orally) during pregnancy and lactation. There are also evidences of depressed maternal behaviour in female exposed [61, 62]. There are also reports by Dessi-Fulgheri et al. [63] about decrease in play behaviour of juvenile Sprague-Dawley rats due to exposure of BPA. Authors observed a masculinization of female behaviour in two behavioural categories, that is, play with females and sociosexual exploration, an effect probably mediated by the estrogenic activity of BPA in the central nervous system.

Foetal/neonatal exposure to low doses of BPA causes sex differences in brain structure, chemistry and behaviour. BPA interferes with the normal processes of sexual differentiation, with brain changes in both male and female rat and mice [61, 64]. Evidence of anatomical alterations in brain sexual differentiation was evident in male and female offspring born to mother exposed to 25 or 250 ng BPA/kg body weight per day [65]. In Fujimoto's experiment, prenatal exposure to BPA affected male rats and abolished sex differences in rearing behaviour in the open-field test and struggling behaviour in the forced swimming test. **Table 3** shows the summary of the experimental results on the behavioural aspects of laboratory rodents.

Event	Model; time and route of exposure	Effect	Citation
Defensive aggression in male	Sprague-Dawley rat, CD-1 mice; developmental, oral	Increased	Farabollini et al. [53]; Kawai et al. [54]
Hyperactivity, hyperlocomotion and rewarding response	Mice, rats; adult, developmental, oral, injection	Increased	Mizuo et al. [56]; Suzuki et al. [57]; Narita et al. [58]; Ishido et al. [59]
Passive and active avoidance learning	Fisher 344 rats; developmental, oral	Impaired	Negishi et al. [60]
Maternal behaviour in females	CD-1 mice, rats; adult, developmental, oral	Decreased	Palanza et al. [61]; Della Seta et al. [62]
Play behaviour in juveniles	Sprague-Dawley rats; developmental, oral	Decreased	Farabollini et al. [63]
Sex differences in behaviour	CD-1 mice, rats; developmental, oral	Lost	Fujimoto et al. [64]; Palanza et al. [61]; Rubin et al. [65]

Table 3. Summary table of the various effects of BPA exposure on behaviour of laboratory rodents.

2.1.5. Other miscellaneous effects

There are evidences on effects of BPA on subsequent activity of enzymes in tissues and thus metabolic processes [66–69]. Study showed very low dose (10 μg/kg) of BPA stimulates insulin production and secretion, which is then followed by insulin resistance at a dose of 100 μg/kg in mice [70]. In the study by Sakurai et al. [71], a high dose of BPA has been revealed to stimulate an increase in the glucose transporter and glucose uptake into adipocytes in cell culture. Study showed that perinatal exposure to a low dose of BPA increased adipogenesis in female rats at weaning [72].

BPA appears to possess complex immuno-modulating effects. It may stimulate or suppress the immune system. It may also alter immune response pathways. There is extensive evidence that BPA modulates both T helper 1 and T helper 2 cytokine production and alters antibody production [73–75]. Yamashita et al. [76] used immune cells from BALB/c mice and demonstrated that BPA induces innate immune response by increasing cytokine synthesis, including tumour necrosis factor (TNF) and IL-1 in macrophages, and stimulates both T and B cells in adaptive response pathway. Using IL-2 and IFN-γ as markers for Th1 response and IL-4 for Th2 response, the authors found that BPA stimulated Th1 cells to produce IFN-γ and Th2 cells to express IL-4. The authors inferred that BPA does not selectively activate the Th1 or Th2 path. BPA also enhances Th1 or Th2 response *in vivo*, depending on the doses [74, 77]. In addition, prenatal exposure to BPA was shown to augment both Th1 and Th2 responses in adulthood [74]. BPA has been reported to modulate immune function at doses between 2.5 and 30 μg/kg/day [70, 73].

2.2. Studies on zebrafish

Zebrafish (*Danio rerio*) as vertebrate model system is popular for studying developmental events. The reasons for choosing zebrafish in developmental biology research include its easy maintenance and rearing, prolific fecundity, transparent embryo, absence of placenta that eases the study of morphological characters and even teratogenic effects on anatomy due to experimental exposure to xenotoxicants. Researchers have taken this opportunity to facilitate their understanding in the effects of BPA on vertebrate model. Summary of the results of experiments on zebrafish model is given in **Table 4**.

2.2.1. *Effects on development and reproduction*

Laboratory studies showed that BPA causes developmental and reproductive effects in zebrafish. There are evidences of delayed hatching, altered axial curvature and tail malformation in zebrafish embryos following exposure of fertilized eggs to BPA [78]. In a study by William et al. [79], BPA altered early dorso-ventral patterning, segmentation and brain development in zebrafish embryos at a concentration of 50 μM within 24 hours of exposure.

Effects on development and reproduction	Endpoint	Life stage and route of exposure	Effect	Citation
	Hatching, axial curvature, tail morphology	Fertilized eggs, directly in a plate	Delayed hatching, altered axial curvature, tail malformation	Hua and Lin [78]
	Early dorso-ventral patterning, segmentation and brain development	Embryo, directly in a plate	Altered	William et al. [79]
	Fertilization and egg production	Breeding adult, in aquarium	Reduced rate of fertilization, increased egg production	Laing et al. [83]
	Testes	Adult, in aquarium	Degenerated, increased number of sustentacular cells, decreased percentage of germ cells	Lora et al. [81]
	Ovary	Adult, in aquarium	Deteriorated ovarian tissues, increased number of atretic follicles, distorted and less developed oocytes	Yon and Akbulut [82]
	Transcription of genes involved in reproductive function	Adult,	Altered	Laing et al. [83]
	Oocyte maturation	Adult, in aquarium	Disrupted	Fitzgerald et al. [84]
Effects on nervous system and behaviour	Hypothalamus	Embryo, directly in culture plate	Increased neurogenesis and hyperactivity	Kinch et al. [89]
	Larval hyperactivity, Adult learning behaviour	Embryo, directly in culture plate	Increased activity, learning deficit	Saili et al. [90]
Effects on chromosomes	Oocyte maturation	Adult, in aquarium	Disrupted by chromatin modification	Santangeli et al. [85]

Table 4. Summary table of the various effects of BPA exposure on zebrafish (*Danio rerio*).

Perturbations in expression of cytochrome P450 aromatase activity have also been observed in zebrafish. Estrogen synthesized in the brain by the action of P450 aromatase is known to have organizing effects on the developing central nervous system. In fish, estrogen increases the predominant brain isoform (P450aromB), implying that xenoestrogens like BPA could act as neurodevelopmental toxicants by altering the expression of P450aromB [80].

Lora et al. [81] found several alterations in the zebrafish testes including a pronounced degeneration of all cellular components, an increase in the percentage of the Sertoli cells and a marked decrease in the percentage of germ cells due to exposure of BPA. Histological studies also showed severe deterioration of ovarian tissue such as disintegration of vesicular structures of mature oocytes, irregularities at cytoplasm, reduction in the number of primary and developing oocytes, deformation at the ooplasm and structure of the mature oocytes and irregularities at nucleolus. The number of the atretic oocytes increased due to BPA exposure. Structurally distorted and less developed oocytes were also observed [82]. A study by Laing et al. [83] documented significant increase in egg production, together with a reduced rate of fertilization in zebrafish exposed to BPA, associated with considerable alterations in the transcription of genes involved in reproductive function and epigenetic processes in both liver (vtg1, esr2b, hdac3, mbd2, mecp2 and dnmt1) and gonad tissue (esr2a, cyp19a1a and amh). Their study demonstrated how BPA disrupts reproductive processes in zebrafish. BPA can also disrupt zebrafish oocyte maturation by a novel nongenomic estrogenic mechanism [84]. BPA exerts this nongenomic estrogenic action on zebrafish oocytes directly through binding to the membrane estrogen receptor Gper and activating a Gper-dependent Egfr/Mapk3/1 pathway. BPA activates this pathway by increasing phosphorylation of Mapk3/1and cAMP concentrations in zebrafish oocytes. Activation of this pathway prevents the resumption of meiotic maturation in fish oocytes [83]. Study showed that BPA downregulated oocyte maturation-promoting signals through changes in the chromatin structure mediated by histone modifications in zebrafish [85].

2.2.2. *Effects on nervous system and behaviour*

Zebrafish has been used extensively to elucidate basic mechanisms underlying behavioural toxicology [86]. Zebrafish was also employed as a model for identifying sex-specific effects on social interactions induced by developmental BPA exposure [87, 88]. A study by Kinch et al. [89] revealed that treatment of embryonic zebrafish with very low-dose BPA (0.0068 μM, 1000-fold lower than the accepted human daily exposure) resulted in 180% increase in neurogenesis within the hypothalamus. Fish embryos exposed to BPA exhibit hyperactivity with ontogenetic growth possibly due to the accelerated neural growth. The authors also found that these effects are probably not due to an effect on estrogen receptors (or estrogen-like receptors) but may be due to its deleterious effects on the synthesis of key enzyme in steroid hormone synthesis, Aromatase B. This study also demonstrated that developmental BPA exposure led to larval hyperactivity or learning deficits in adult zebrafish [90]. There are evidences for temperature-specific impairment of swimming performance, disturbances in muscle activity and gene expression in zebrafish due to exposure of BPA [91]. This result suggests that BPA toxicity is compounded with the effects of climate change.

2.2.3. *Other miscellaneous effects*

BPA can alter sex ratio of zebrafish by inducing feminization of the fry [92]. Zebrafish embryos exposed to BPA also showed signs of feminized brains [86]. Kinch et al. [93] investigated morphological changes to developing zebrafish caused by exposure to BPA including changes in body length, pericardia (heart) and the head. Na et al. [94] observed a significant damage in the liver of zebrafish after 96 hours of exposure to BPA. This result further confirmed that liver was the target organ of BPA.

3. Invertebrate model animals

3.1. Study on *Drosophila melanogaster*

Drosophila melanogaster remains as one of the popular organism in studying the effects of BPA on eukaryotic biological system. The study on *Drosophila* includes change in gene expression profile, change in behaviour and nervous system, alteration in juvenile growth and development, history traits and fecundity and metabolism.

3.1.1. *Effects on life history traits and developmental event*

In comparison to other studies on effects of BPA on biological aspects in *Drosophila melanogaster*, adequate references are available on the researches on *Drosophila* life history traits. The effects of BPA on growth and development in *Drosophila* were observed, which demonstrated a statistically significant increase in larval growth for the low-dose treatment group (0.1 mg/L), but not in the high-dose treatment group (10 mg/L). BPA exposure caused an increase in body size in treated flies at 48, 72 and 96 hours following egg laying (AEL), suggesting a non-monotonic dose response. The increase in growth rate found for all treatment groups was associated with a statistically significant increase in food intake observed at 72-hour AEL. Furthermore, it was observed that the increased growth rate was coupled with an earlier onset of pupariation and metamorphosis, resulting from increased activity of insulin/insulin growth factor signalling (IIS) in *Drosophila*. Thus, this suggests that BPA exerts its effects through disruption of endocrine signalling in *Drosophila* since the timing of the onset of pupariation in *Drosophila* is controlled through the complex interaction of the IIS and the ecdysone signalling pathways. All these observations suggest that the effect is probably due to disruption of insulin-like signalling in cellular system [95].

Another study on life history traits of *Drosophila* [96] obtained some contradiction to the above-mentioned observation. The author reported a delay in both the mean pupation and the mean maturation times in treated group. In that experiment, larvae of *D. melanogaster* were exposed to three different concentrations: 0.1, 1 and 10 mg/L BPA. In the 0.1 and 1 mg/L exposed groups, the mean offspring numbers were significantly less than that of the control groups, indicating that mean fecundity was significantly decreased. Thus, administration of BPA in both food and through body wall absorption resulted in altered fecundity [96]. Mean decrease in fecundity as compared to control in *Drosophila* exposed to BPA is also evident in the work of Atli et al. [96]. William et al. [97] have reported that BPA exposure causes inhibition of lipolysis during starvation, leading to significantly increased lipid content after 24 hours of fasting. Furthermore, it also suppresses the expression of insulin-like peptide in *Drosophila,* indicating that BPA may inhibit lipid recruitment during starvation in *Drosophila.*

3.1.2. Effects on behaviour and nervous system

BPA causes [98] behavioural modifications in *Drosophila melanogaster,* which, in turn, suggests intuitively the role of environmental risk factors for the behavioural impairments like autism and attention deficit hyperactivity disorder (ADHD) in human. The study revealed disturbance in the locomotion patterns of BPA-exposed *Drosophila* that may relate to the decision-making and the motivational state of the animal. Furthermore, an increase in repetitive behaviour and disturbance in grooming behaviour and abnormal social interaction of *Drosophila* following BPA exposure were seen.

A recent study conducted by Streifel [99] shows that administration of BPA in the prenatal environment had significant impacts on some aspects of *Drosophila* behaviour, which includes increased time spent in seeking behaviour, increased numbers of peristaltic contractions, increased linear as well as angular movement, decrease in turn angle value as well as potentially significant impacts on motor nerve morphology. These findings suggest implication of BPA as ubiquitous neurotoxin that acts upon the delicate process of neurodevelopment.

3.1.3. Effects on global gene expression profile

Alteration in gene expression profile in *Drosophila* has been studied by Branco et al. [100]. The authors reported that the effects due to BPA on genome-wide gene expression of *D. melanogaster* can be enhanced by the ingestion of high dietary sugar. The authors have found that acute and chronic exposure to BPA causes gross downfall in transcription of testis-specific genes and overexpression of ribosome-associated genes across tissues. In addition, it causes alteration of transposable elements that are specific to the ribosomal DNA loci, suggesting that nucleolar stress might implicate in BPA toxicity. This observation suggests that BPA and dietary sugar might functionally interact, with consequences to regulatory programmes in both reproductive and somatic tissues [100].

3.2. Study on other invertebrate model

As compared to vertebrates, the number of research works regarding BPA exposure on invertebrates is minimum. Invertebrates are frequently used as bioindicators for endocrine-disrupting chemicals. Research suggests that some invertebrates appear to be quite sensitive to BPA, and effects have been documented even at environmentally relevant concentrations [101].

3.2.1. Effects on life history traits and developmental events

A study conducted by Lemos et al. [102] revealed that low BPA concentrations disrupt the endocrine function of terrestrial arthropod *Porcellio scaber* by causing a sex-ratio shift. In this study, endocrine system-related chronic effects were identified at a lower dose of BPA than the concentration having acute toxic effects on isopods, indicating impairment of molting, incomplete ecdysis.

The effects of various concentrations of BPA on the development of two sea urchin species *Hemicentrotus pulcherrimus* and *Strongylocentrotus nudus* were examined [103]. This study suggested that the sensitivity of sea urchin embryos and juveniles to endocrine disrupter chemicals changes during the stages of development. The development in the first 12 hours following fertilization up to the morphogenesis of embryo was found to be most sensitive. Even higher concentrations of BPA exposure (>300 mg/L) resulted in developmental arrest and mortality in the sea urchin *Paracentrotus lividus* [104].

Studies on lepidopteran corn stalk borer *Sesamia nonagrioides* revealed that BPA induces various developmental disorders through interfering effect in ecdysteroidal pathway [105] and over expression of heat-shock proteins [106]. Study on freshwater insect *Chironomus riparius* showed that adult emergence times were significantly delayed on moderate BPA exposure [107]. Marcial et al. (2003) and Watts et al. [108, 109] found that the marine copepod *Tigriopus japonicus* showed developmental inhibition at a very low concentration of BPA (0.1 mg/L). However, it is unclear if these effects have any long-term impacts in adult life. Experimental exposure to higher concentration of (11.4 mg/L) BPA for 1 hour caused premature larval metamorphosis in the marine polychaete worm *Capitella capitata* [110].

A study conducted on *Hydra vulgaris* by Pascoe et al. [111] pointed that the structure and physiology of polyps were adversely affected at concentrations greater than 42 μg/L BPA. Also, inhibition of regeneration ability was recorded above 460 μg/L BPA concentration. The results indicate that signalling processes necessary for the control and regulation of cell movement and differentiation during normal development, regeneration and sexual reproduction in *H. vulgaris* are not disrupted by BPA at low environmentally relevant concentrations.

3.2.2. Effects on reproductive system and fecundity

As far as published literatures are concerned, several studies have been conducted to unravel the adverse effects of BPA on reproductive systems and reproductive functioning in various invertebrate animals. In the study of Manshilha et al. [112], an increased fecundity (neonates per female), in comparison with the negative control group (100.3 ± 1.6%), was observed when daphnids were cultured and allowed to breed in the polycarbonate (PC) containers (145.1 ± 4.3%–264.7 ± 3.8%) for single and multiple generations. A strong dose-dependent ecotoxicological effect was evident, and it was suggested that BPA leached from plastic materials acts as functional estrogen *in vivo* at very low concentrations. In contrast, neonate production by daphnids cultured in polypropylene and non-PC bottles was slightly but not significantly enhanced (92.5 ± 2.0 to 118.8 ± 1.8%). Multigenerational tests also demonstrated magnification of the adverse effects, not only on fecundity but also on mortality of the species. Reproductive impairment in Daphnia due to exposure to BPA is also evident in the study by Tišler et al. [113].

Andersen et al. [6] found an increase in egg production in copepod *Acartiatonsa* exposed to 20 μg BPA/L. Moreover, inhibition in normal development at BPA concentrations above environmentally relevant levels (100 mg/L) was also evident. At extremely high exposures (16,000–80,000 mg/L), abnormal growth and inhibition of gemule germination was found in freshwater sponges *Heteromyenia sp.* and *Eunapius fragilis* [114]. A study conducted by Oehlmann et al. [115] on freshwater snail *Marisa cornuarietis* and of the marine prosobranch *Nucella lapillus* revealed that BPA affects the reproductive system and has a negative impact on snails even at nominal concentration, that is, 1 μg/L. Affected *Marisa* females were designated as 'superfemales' and were characterized by the presence of additional female organs, hyperplasia of the accessory pallial sex glands, malformations of the pallial oviduct causing increased female mortality and a strong stimulation of oocyte and spawning mass production. In these follow-up studies, Oehlmann et al. [116] tried to bridge several gaps in knowledge by conducting additional experiments. Here, the authors confirm the previous results and additionally conclude that the occurrence of superfemales is associated with adverse effects on reproduction and survival, even at sub-micrograms per litre concentrations of BPA (NOEC, 7.9 ng/L; EC10, 13.9 ng/L). However, if snails are exposed to BPA under conditions that maximize the reproductive output, particularly during the spawning season or at elevated temperatures, the induction of superfemales is at least partially masked. The superfemale induction is probably mediated by binding of BPA with estrogen receptor, because the response can completely be reversed by coexposure to potent estrogen inhibitors. Furthermore, the extreme BPA sensitivity of *M. cornuarietis* and other prosobranch snails probably due to higher affinity of the compound for the estrogen receptor in this species was compared. Overall, the results suggest that BPA imposes a potential hazard for prosobranch population in the field even at environmentally relevant concentrations. Experimentally determined EC_{50} values of BPA for different invertebrate model organisms have been given in **Table 5**.

3.2.3. *Effects on gene expression profile*

Change in expression pattern of genes and alteration in RNA expression pattern due to BPA exposure are also within the scientific interest. Planelló et al. [117] studied the effects of BPA on the expression of some selected genes, including housekeeping, stress-induced and hormone-related genes in *C. riparius* larvae. They found that exposure to BPA at a concentration of 3 mg/L for 12–24-hour exposure did not influence the levels of ribosomal RNA or those

Species	EC_{50} (mg/L)	NOEC (mg/L)	Reference
Waterflea Daphnia magna	10.2	4.1	Alexander et al. [4]
Mysid Mysidopsis bahia	1.1	0.51	Surprenant [123]
Chironomid Chironomus tentans	2.7	1.4	Mihaich et al. [124]
Copepod Tigriopus japonicus	4.32	3.5	Marcial et al. [108]
Snail Marisa cornuarietis	>4.03 (LC_{50})	1.32	Mihaich et al. [124]
Snail Marisa cornuarietis	2.24 (LC_{50})	1.18	Mihaich et al. [124]

Table 5. Summarized presentation showing experimentally determined effective concentration (EC_{50}) and no effect concentration (NOEC) of BPA on different invertebrate animals [103].

of mRNAs for both L11 or L13 ribosomal proteins which were selected as representative of housekeeping genes involved in ribosome biogenesis. Nonetheless, BPA treatment induced the transcription of the HSP70 gene. Interestingly, BPA causes significant increase in transcript of the ecdysone receptor (EcR), suggesting that BPA can selectively affect the expression of the ecdysone receptor gene suggesting a direct interaction with the insect endocrine system.

Significant level of DNA strand break has been detected in snail *Potamopyrgus antipodarum* under exposure to BPA [118]. DNA-damaging effect of BPA on aquatic insect *C. riparius* has also been reported by Martinez-Paz et al. [119].

4. Conclusion

Bisphenol-A (BPA), found ubiquitously in our environment, has received a tremendous amount of attention from research scientists, government panels and the popular press. Extensive investigational work has been and is still being carried out in various fields like: (1) mechanisms of BPA action; (2) levels of human exposure; (3) routes of human exposure; (4) pharmacokinetic models of BPA metabolism; (5) effects of BPA on exposed animals and (6) links between BPA and cancer. BPA interferes with hormone signalling via two mechanisms: altering the availability of ovarian hormones and altering binding and activity of the hormone at the receptor level [120–122].

Besides understanding the probable human health hazards, study of BPA effect on model organisms facilitates our concern to the issues like biodiversity loss, environmental degradation and overall imbalance in ecological functioning. Today's world is extremely dependent on plastics, and this dependency inevitably brings the challenges of BPA exposure to the environment. Invertebrate and vertebrate fauna from terrestrial and aquatic ecosystems get affected equally, and the situation is going worse every day. Tantalizingly, the role of BPA in biodiversity loss is not being analysed when the issue comes on the table for discussion. So, mass awareness is to be build up among the people that include students, scholar, academician, conservationist, wildlife activist, NGOs working with environmental issues, policy-makers and politicians across the nation. It is hard to make BPA free world, but the extent of its adverse effect could be mitigated by our concern and consciousness.

Author details

Papiya Ghosh[1], Sohini Singha Roy[2], Morium Begum[2] and Sujay Ghosh[2*]

*Address all correspondence to: sgzoo@caluniv.ac.in

1 Department of Zoology, Bijoykrishna Girls' College, Howrah, India

2 Department of Zoology, University of Calcutta, Kolkata, India

References

[1] Bhandari RK, Deem SL, Holliday DK, Jandegian CM, Kassotis CD, Nagel SCH, Tillitt DE, vom Saal FS, Rosenfeld CS. Effects of the environmental estrogenic contaminants bisphenol A and 17α-ethinyl estradiol on sexual development and adult behaviors in aquatic wildlife species. General and Comparative Endocrinology. 2015;**214**:195–219

[2] Yoon K, Kwack SJ, Kim HS, Lee BM. Estrogenic endocrine-disrupting chemicals: Molecular mechanisms of actions on putative human diseases. Journal of Toxicology and Environmental Health Part B Critical Reviews. 2014;**17**(3):127–174

[3] Fenichel P, Chevalier N, Brucker-Davis F. Bisphenol A: An endocrine and metabolic disruptor. Annales D Endocrinologie (Paris). 2013;**74**(3):211–220

[4] Alexander HC, Dill DC, Smith LW, Guiney PD, Dorn P. Bisphenol A: Acute aquatic toxicity. Environmental Toxicology and Chemistry. 1988;**7**:19–26

[5] Sohoni P, Tyler CR, Hurd K, Caunter J, Hetheridge M, Williams T, Woods C, Evans M, Toy R, Gargas M, Sumpter JP. Reproductive effects of long-term exposure to bisphenol A in the fathead minnow (*Pimephales promelas*). Environmental Science & Technology. 2001;**35**:2917–2925

[6] Andersen HR, Halling-Sorensen B, Kusk KO. A parameter for detecting estrogenic exposure in the copepod *Acartia tonsa*. Ecotoxicology and Environmental Safety. 1999;**44**:56–61

[7] Pellegrini M, Bulzomi P, Lecis M, Leone S, Campesi I, Franconi F, Marino M. Endocrine disruptors differently influence estrogen receptor β and androgen receptor in male and female rat VSMC. Journal of Cellular Physiology. 2014;**229**(8):1061–1068

[8] Matthews JB, Twomey K, Zacharewski TR. In vitro and in vivo interactions of bisphenol A and its metabolite, bisphenol A glucuronide, withestrogen receptors alpha and beta. Chemical Research in Toxicology. 2001;**14**:149–157

[9] Snyder RW, Maness SC, Gaido KW, Sumner SCJ, Fennell TR. Metabolism and disposition of bisphenol A in female rats. Toxicology and Applied Pharmacology. 2000;**168**:225–234

[10] Volkel W, Colnot T, Csanady GA, Filser JG, Dekant W. Metabolism and kinetics of bisphenol A in humans at low doses following oral administration. Chemical Research in Toxicology. 2002;**15**:1281–1287

[11] Volkel W, Bittner N, Dekant W. Quantitation of bisphenol A and bisphenol A glucuronide in biological samples by high performance liquid chromatography–tandem mass spectrometry. Drug Metabolism and Disposition. 2005;**33**:1748–1757

[12] Susiarjo M, Hassold TJ, Freeman E, Hunt PA. Bisphenol A exposure *in utero* disrupts early oogenesis in the mouse. PLoS Genetics. 2007;**3**:63–70

[13] Soto AM, Vandenberg LN, Maffini MV, Sonnenschein C. Does breast cancer start in the womb? Basic & Clinical Pharmacology & Toxicology. 2008;**102**:125–133

[14] Newbold RR, Jefferson WN, Padilla-Banks E. Long-term adverse effects of neonatal exposure to bisphenol A on the murine female reproductive tract. Reproductive Toxicology. 2007;**24**:253–258

[15] Manfo FP, Jubendradass R, Nantia EA, Moundipa PF, Mathur PP. Adverse effects of bisphenol A on male reproductive function. Reviews of Environmental Contamination and Toxicology. 2014;**228**:57–82

[16] Akingbemi BT, Sottas CM, Koulova AI, Klinefelter GR, Hardy MP. Inhibition of testicular steroidogenesis by the xenoestrogenbisphenol A is associated with reduced pituitary luteinizing hormone secretion and decreased steroidogenic enzyme gene expression in rat Leydig cells. Endocrinology. 2004;**145**(2):592–603

[17] Aikawa H, Koyama S, Matsuda M, Nakahashi K, Akazome Y, Mori T. Relief effect of vitamin A on the decreased motility of sperm and the increased incidence of malformed sperm in mice exposed neonatally to bisphenol A. Cell and Tissue Research. 2004;**315**(1):119–124

[18] Toyama Y, Suzuki-Toyota F, Maekawa M, Ito C, Toshimori K. Adverse effects of bisphenol A to spermiogenesis in mice and rats. Archives of Histology and Cytology. 2004;**67**(4): 373–381

[19] vomSaal FS, Cooke PS, Buchanan DL, Palanza P, Thayer KA, Nagel SC, et al. A physiologically based approach to the study of bisphenol A and other estrogenic chemicals on the size of reproductive organs, daily sperm production, and behavior. Toxicology and Industrial Health. 1998;**14**(12):239–260

[20] Timms BG, Howdeshell KL, Barton L, Bradley S, Richter CA, vomSaal FS. Estrogenic chemicals in plastic and oral contraceptives disrupt development of the mouse prostate and urethra. Proceedings of the National Academy of Sciences of the United States of America. 2005;**102**(19):7014–7019

[21] Lagos-Cabré R, Moreno RD. Contribution of environmental pollutants to male infertility: A working model of germ cell apoptosis induced by plasticizers. Biological Research. 2012;**45**(1):5–14

[22] Wong EW, Cheng CY. Impacts of environmental toxicants on male reproductive dysfunction. Trends in Pharmacological Sciences. 2011;**32**(5):290–299

[23] Al-Hiyasat AS, Darmani H, Elbetieha AM. Effects of bisphenol A on adult male mouse fertility. European Journal of Oral Sciences. 2002;**110**(2):163–167

[24] Chitra KC, Latchoumycandane C, Mathur PP. Induction of oxidative stress by bisphenol A in the epididymal sperm of rats. Toxicology. 2003;**185**(12):119–127

[25] Takao T, Nanamiya W, Nagano I, Asaba K, Kawabata K, Hashimoto K. Exposure with the environmental estrogen bisphenol A disrupts the male reproductive tract in young mice. Life Science. 1999;**65**(22):2351–2357

[26] Sakaue M, Ohsako S, Ishimura R, Kurosawa S, Kurohmaru M, Hayashi Y, et al. Bisphenol A affects spermatogenesis in the adult rat even at a low dose. Journal of Occupational Health. 2001;**43**:185–190

[27] Thayer KA, Ruhlen RL, Howdeshell KL, Buchanan DL, Cooke PS, Preziosi D, et al. Altered prostate growth and daily sperm production in male mice exposed prenatally to subclinical doses of 17aethinyl oestradiol. Human Reproduction. 2001;**16**(5):988–996

[28] Jeng HA. Exposure to endocrine disrupting chemicals and male reproductive health. Frontiers in Public Health. 2014;**5**;2:55

[29] Tohei A, Suda S, Taya K, Hashimoto T, Kogo H. Bisphenol A inhibits testicular functions and increases luteinizing hormone secretion in adult male rats. ExpBiol Med 2001;**226**:216–221

[30] Palioura E, Diamanti-Kandarakis E. Industrial endocrine disruptors and polycystic ovary syndrome. Journal of Endocrinological Investigation. 2013;**36**(11):1105–1111

[31] Benagiano G, Brosens I. In utero exposure and endometriosis. The Journal of Maternal Fetal & Neonatal Medicine. 2014;**27**(3):303–308

[32] Markey CM, Luque EH, Munoz De Toro M, Sonnenschein C, Soto AM. In utero exposure to bisphenol A alters the development and tissue organization of the mouse mammary gland. Biology of Reproduction. 2001;**65**(4):1215–1223

[33] Munoz-de-Toro M, Markey CM, Wadia PR, Luque EH, Rubin BS, Sonnenschein C, et al. Perinatal exposure to bisphenol-A alters peripubertal mammary gland development in mice. Endocrinology. 2005;**146**(9):4138–4147

[34] Durando M, Kass L, Piva J, Sonnenschein C, Soto AM, Luque E, et al. Prenatal bisphenol A exposure induces preneoplastic lesions in the mammary gland in Wistar rats. Environmental Health Perspectives. 2007;**115**(1):80–86

[35] Murray TJ, Maffini MV, Ucci AA, Sonnenschein C, Soto AM. Induction of mammary gland ductal hyperplasias and carcinoma *in situ* following fetal bisphenolA exposure. Reproductive toxicology (Elmsford, NY). 2007;**23**(3):383–390. DOI: 10.1016/j.reprotox.2006.10.002

[36] Simerly RB. Wired for reproduction: Organization and development of sexually dimorphic circuits in the mammalian forebrain. Annual Review of Neuroscience. 2002;**25**: 507–536

[37] Itoh K, Yaoi T, Fushiki S. Bisphenol A, an endocrine-disrupting chemical, and brain development. Neuropathology. 2012;**32**(4):447–457

[38] Negri-Cesi P. Bisphenol A interaction with brain development and functions. Dose-Response. 2015;**13**(2):1559325815590394. DOI: 10.1177/1559325815590394

[39] Porterfield SP, Hendrich CE. The role of thyroid hormones in prenatal and neonatal neurological development—Current perspectives. Endocrine Reviews. 1993;**14**:94–106

[40] Schantz SL, Widholm JJ. Cognitive effects of endocrine-disrupting chemicals in animals. Environmental Health Perspectives. 2001;**109**:1197–1206

[41] Nakamura K, Itoh K, et al. Murine neocortical histogenesis is perturbed by prenatal exposure to low doses of Bisphenol A. Journal of Neuroscience Research. 2006;**84**(6):1197–1205

[42] MacLusky NJ, Hajszan T, Leranth C. The environmental estrogen bisphenol A inhibits estrogen induced hippocampal synaptogenesis. Environmental Health Perspectives. 2005;**113**:675–679

[43] Leranth C, Szigeti-Buck K, Maclusky NJ, Hajszan T. Bisphenol A prevents the synaptogenic response to testosterone in the brain of adult male rats. Endocrinology. 2007;**149**:988–994

[44] Obata T, Kubota S. Formation of hydroxy radicals by environmental estrogen-like chemicals in rat striatum. Neuroscience Letters. 2000;**296**:41–44

[45] Obata T. Imidaprilat, an angiotensin-converting enzyme inhibitor exerts neuroprotective effect via decreasing dopamine efflux and hydroxyl radical generation induced by bisphenol A and MPP+ in rat striatum. Brain Research. 2006;**1071**:250–253

[46] Lin Y, Zeng XG, Wu DS, Wang X. Study on bisphenol A induced primary cultured mesencephalicneuronal cell injury by oxidative stress. Wei Sheng Yan Jiu. 2006;**35**:419–422

[47] Hunt PA, Koehler KE, Susiarjo M, Hodges CA, Hagan A, Voigt RC, et al. Bisphenol A causes meiotic aneuploidy in the female mouse. Current Biology. 2003;**13**:546–553

[48] Tsutsui T, Tamura Y, Yagi E, Hasegawa K, Takahashi M, Maizumi N, et al. Bisphenol-A induces cellular transformation, aneuploidy and DNA adduct formation in cultured Syrian hamster embryo cells. International Journal of Cancer. 1998;**75**(2):290–294

[49] Tsutsui T, Tamura Y, Suzuki A, Hirose Y, Kobayashi M, Nishimura H, et al. Mammalian cell transformation and aneuploidy induced by five bisphenols. International Journal of Cancer. 2000;**86**(2):151–154

[50] Parry EM, Parry JM, Corso C, Doherty A, Haddad F, Hermine TF, et al. Detection and characterization of mechanisms of action of aneugenic chemicals. Mutagenesis. 2002;**17**(6):509–521

[51] Eichenlaub-Ritter U, Sun F, Betzendahl I. Meiotic progression, spindle formation and chromosome segregation in in vitro maturing mouse oocytes exposed to bisphenol A. Environmental Research. 2005;**98**(3):405–406

[52] Can A, Semiz O, Cinar O. Bisphenol-A induces cell cycle delay and alters centrosome and spindle microtubular organization in oocytes during meiosis. Molecular Human Reproduction. 2005;**11**(6):389–396. DOI: 10.1093/molehr/gah179

[53] Farabollini F, Porrini S, Della Seta D, Bianchi F, Dessi-Fulgheri F. Effects of perinatal exposure to bisphenol A on sociosexual behavior of female and male rats. Environmental Health Perspectives. 2002;**110**(Suppl 3):409–414

[54] Kawai K, Takehiro N, Nishikata H, Aou S, Takii M, Kubo C. Aggressive behavior and serum testosterone concentration during the maturation process of male mice: The effects of fetal exposure to bisphenol A. Environmental Health Perspectives. 2003;**111**:175–178

[55] Kajta M, Wójtowicz AK. Impact of endocrine-disrupting chemicals on neural development and the onset of neurological disorders. Pharmacological Reports. 2013;**65**(6):1632–1639

[56] Mizuo K, Narita M, Miyagawa K, Okuno E, Suzuki T. Prenatal and neonatal exposure to bisphenolA affects the morphine-induced rewarding effect and hyperlocomotion in mice. Neuroscience Letters. 2004;**356**(2):95–98

[57] Suzuki T, Mizuo K, Nakazawa H, Funae Y, Fushiki S, Fukushima S, Shirai T, Narita M. Prenatal and neonatal exposure to bisphenol-A enhances the central dopamine D1 receptor-mediated action in mice: Enhancement of the methamphetamine-induced abuse state. Neuroscience. 2003;**117**:639–644

[58] Narita M, Miyagawa K, Mizuo K, Yoshida T, Suzuki T. Changes in central dopaminergic systems and morphine reward by prenatal and neonatal exposure to bisphenol-A in mice: Evidence for the importance of exposure period. Addiction Biology. 2007;**12**:167–172

[59] Ishido M, Morita M, Oka S, Masuo Y. Alteration of gene expression of G protein coupled receptors in endocrine disruptors-caused hyperactive rats. Regulatory Peptides. 2005;**126**:145–153

[60] Negishi T, Kawasaki K, Suzaki S, Maeda H, Ishii Y, Kyuwa S, et al. Behavioral alterations in response to fear-provoking stimuli and tranylcypromine induced by perinatal exposure to bisphenol A and nonylphenol in male rats. Environ Health Perspect 2004;**112**(11):1159–1164

[61] Palanza PL, Gioiosa L, Parmigiani S, vomSaal FS. Effects of developmental exposure to bisphenolA on brain and behavior in mice. Environmental Research. 2008;**108**:150–157

[62] Della Seta D, Minder I, Dessi-Fulgheri F, Farabollini F. Bisphenol-A exposure during pregnancy and lactation affects maternal behavior in rats. Brain Res Bull 2005;**65**(3):255–260

[63] Dessi-Fulgheri F, Porrini S, Farabollini F. Effects of perrinatal exposure to bisphenol A on play behavior of female and male juvenile rats. Environ Health Perspect 110 Suppl 2002;**3**:403-407

[64] Fujimoto T, Kubo K, Aou S. Prenatal exposure to bisphenol A impairs sexual differentiation of exploratory behavior and increases depression-like behavior in rats. Brain Research. 2006;**1068**:49–55

[65] Rubin BS, Lenkowski JR, Schaeberle CM, Vandenberg LN, Ronsheim PM, Soto AM. Evidence of altered brain sexual differentiation in mice exposed perinatally to low, environmentally relevant levels of bisphenol A. Endocrinology. 2006;**147**:3681–3691

[66] Heine PA, Taylor JA, Iwamoto GA, Lubahn DB, Cooke PS. Increased adipose tissue in male and female estrogen receptor-α knockout mice. Proceedings of the National Academy of Sciences of the United States of America. 2000;**97**(23):12729–12734

[67] Cooke PS, Heine PA, Taylor JA, Lubahn DB. The role of estrogen and estrogen receptor-alpha in male adipose tissue. Molecular and Cellular Endocrinology. 2001;**178**(12):147–154

[68] Kabuto H, Hasuike S, Minagawa N, Shishibori T. Effects of bisphenol A on the metabolisms of active oxygen species in mouse tissues. Environmental Research. 2003;**93**(1):31–35

[69] Alonso-Magdalena P, Morimoto S, Ripoll C, Fuentes E, Nadal A. The estrogenic effect of bisphenol A disrupts pancreatic β-cell function *in vivo* and induces insulin resistance. Environmental Health Perspectives. 2006;**114**(1):106–112

[70] Ropero AB, Alonso-Magdalena P, Garcia-Garcia E, Ripoll C, Fuentes E, Nadal A. Bisphenol-A disruption of the endocrine pancreas and blood glucose homeostasis. International Journal of Andrology. 2008;**31**:194–200

[71] Sakurai K, Kawazuma M, Adachi T, Harigaya T, Saito Y, Hashimoto N, et al. Bisphenol A affects glucose transport in mouse 3T3-F442A adipocytes. Br J Pharmacol. 2004;**141**(2):209–214

[72] Somm E, Schwitzgebel VM, Toulotte A, Cederroth CR, Combescure C, Nef S, Aubert ML, Huppi PS. Perinatal exposure to bisphenol a alters early adipogenesis in the rat. Environmental Health Perspectives. 2009;**117**:1549–1555

[73] Sawai C, Anderson K, Walser-Kuntz D. Effect of bisphenolA on murine immune function: Modification of interferon-γ, IgG2a, and disease symptoms in NZB × NZW F1 mice. Environmental Health Perspectives. 2003;**111**(16):1883–1887

[74] Yoshino S, Yamaki K, Yanagisawa R, Takano H, Hayashi H, Mori Y. Effects of bisphenol A on antigen-specific antibody production, proliferative responses of lymphoid cells, and TH1 and TH2 immune responses in mice. British Journal of Pharmacology. 2003;**138**(7):1271–1276

[75] Ohshima Y, Yamada A, Tokuriki S, Yasutomi M, Omata N, Mayumi M. Transmaternal exposure to bisphenol a modulates the development of oral tolerance. Pediatric Research. 2007;**62**:60–64

[76] Yamashita U, Sugiura T and Kuroda E. Effect of endocrine disrupters on immune responses in vitro. J UOEH 2002;**24**:1–10

[77] Tian X, Takamoto M, Sugane K. Bisphenol A promotes IL-4 production by Th2 cells. International Archives of Allergy and Immunology. 2003;**132**:240–247

[78] Hua DZ, Lin Z. Toxicity of bisphenol A on the growth of zebrafish embryos. Acta Hydrobiologica Sinica. 2006;**30**(6):638–642

[79] William KF, Yeung BHY, Wan HT, Wong CKC. Early embryogenesis in zebrafish is affected by bisphenolAexposure. Biology Open 2013;**2**(5):466-471. DOI:10.1242/bio.20134283

[80] Kishida M, McLellan M, Miranda JA, Callard GV. Estrogen and xenoestrogens upregulate the brain aromatase isoform (P450aromB) and perturb markers of early 39 development in zebrafish (*Danio rerio*). Comparative Biochemistry and Physiology Part B Biochemistry & Molecular Biology. 2001;**129**:261–268

[81] Lora AJ, Molina AM, Bellido C, Blanco A, Monterde JG, Moyano MR. Adverse effects of bisphenol A on the testicular parenchyma of zebrafish revealed using histomorphological methods. VeterinarniMedicina 2016;**61**(10):577–589 DOI: 10.17221/212/2015-VETMED

[82] Yön ND, Akbulut C. Histological changes in zebrafish (*Danio rerio*) ovaries following administration of bisphenol A. Pakistan Journal of Zoology. 2014;**46**(4):1153–1159

[83] Laing LV, Viana J, Dempster EL, et al. Bisphenol A causes reproductive toxicity, decreases *dnmt1* transcription, and reduces global DNA methylation in breeding zebrafish *(Danio rerio)*. Epigenetics. 2016;**11**(7):526–538. DOI: 10.1080/15592294.2016.1182272

[84] Fitzgerald AC, Peyton C, Dong J, Thomas P. Bisphenol A and related alkylphenols exert nongenomic estrogenic actions through a G protein-coupled estrogen receptor 1 (GPER)/epidermal growth factor receptor (EGFR) pathway to inhibit meiotic maturation of zebrafish oocytes. Biology of Reproduction. 2015;**93**(6):135. DOI: 10.1095/biolreprod.115.132316

[85] Santangeli S, Maradonna F, Gioacchini G, Cobellis G, Piccinetti CC, Dalla Valle L, Carnevali O. BPA-Induced deregulation of epigenetic patterns: Effects on female zebrafish reproduction. Scientific Reports. 2016;**6**:21982. DOI: http://doi.org/10.1038/srep21982

[86] Bailey J, Oliveri A, Levin ED. Zebrafish model systems for developmental neurobehavioral toxicology. Birth Defects Research Part C Embryo Today. 2013;**99**:14–23

[87] Dahlbom SJ, Backström T, Lundstedt-Enkel K, Winberg S. Aggression and monoamines: Effects of sex and social rank in zebrafish (*Danio rerio*). Behavioural Brain Research. 2012;**228**:333–338

[88] Weber DN, Hoffmann RG, Hoke ES, Tanguay RL. Bisphenol A exposure during early development induces Sex-Specific changes in adult zebrafish social interactions. Journal of Toxicology and Environmental Health Part A. 2015;**78**(1):50–66. DOI: 10.1080/15287394.2015.958419

[89] "https://www.ncbi.nlm.nih.gov/pubmed/?term=Kinch%20CD%5BAuthor%5D&cauthor=true&cauthor_uid=27107150" Kinch CD, "https://www.ncbi.nlm.nih.gov/pubmed/?term=Kurrasch%20DM%5BAuthor%5D&cauthor=true&cauthor_uid=27107150" Kurrasch DM, "https://www.ncbi.nlm.nih.gov/pubmed/?term=Habibi%20HR%5BAuthor%5D&cauthor=true&cauthor_uid=27107150" Habibi HR. Adverse morphological development in embryonic zebrafish exposed to environmental concentrations of contaminants individually and in mixture. Aquat Toxicol 2016;**175**:286–298. doi: 10.1016/j.aquatox.2016.03.021.

[90] Saili KS, Corvi MM, Weber DN, Patel AU, Das SR, Przybyla J, Anderson KA, Tanguay RL. Neurodevelopmental low-dose bisphenolA exposure leads to early life-stage hyperactivity and learning deficits in adult zebrafish. Toxicology. 2012;**291**:83–92

[91] Little AG, SeebacherF. Temperature determines toxicity: Bisphenol A reduces thermal tolerance in fish. Environmental Pollution. 2015;**197**:84–89

[92] Drastichová J, Svobodová Z, Groenland M, Dobsíková R, Îlábek V, Weissová D, Szotkowská M. Effect of exposure to bisphenol A and 17β-estradiol on the sex differentiation in zebrafish (*Danio rerio*). Acta Veterinaria Brno. 2005;**74**:287–291

[93] Kinch CD, Kurrasch DM, Habibi HR. Adverse morphological development in embryonic zebra fish exposed to environmental concentration of contaminants individually and in mixture. Aquat Toxicol 2015;**175**:286–298

[94] Na CAO, Hua WEI, Ling-guang WU, Ting-ting WU, Guo-peng LI. Effects of bisphenol A on zebrafish (*Daniorerio*) liver and gonad.Chinese Journal of Ecology 2010;**29**(11):2192–2198

[95] Weiner AK, Ramirez A, Zintel T, Rose RW, Wolff E, Parker AL, Bennett K, Johndreau K, Rachfalski C, Zhou J, Smith ST. Bisphenol A affects larval growth and advances the onset of metamorphosis in *Drosophila melanogaster*. Ecotoxicology and Environmental Safety. 2014;**101**:7–13

[96] Atli E, Ünlü H. Developmental and reproductive effects of bisphenol A (Bpa) in *Drosophila melanogaster*. The Journal of Biological Chemistry. 2012;**40**(1):61–68

[97] Williams MJ, Wang Y, Klockars A, Monica Lind P, Fredriksson R, Schioth HB. Exposure to Bisphenol A affects lipid metabolism in Drosophila melanogaster. Basic Clin Pharmacol Toxicol 2014;**114**(5):414–420

[98] Kaur K, Simon AF, Chauhan V, Chauhan A. Effect of bisphenol A on *Drosophila melanogaster* behavior—A new model for the studies on neurodevelopmental disorders. Behavioural Brain Research. 2015;**284**:77–84

[99] Streifel A M. Effect of bisphenol-A on neurodevelopment in *Drosophila melanogaster* larvae.2016. (Thesis)

[100] Branco AT, Lemos B. Interaction between bisphenol A and dietary sugar affects global gene transcription in *Drosophila melanogaster*. Genomics Data. 2014;**2**:308–311

[101] Oehlmann J, Schulte-Oehlmann U, Kloas W, Jagnytsch O, Lutz I, Kusk KO, Wollenberger L, Santos EM, Paull GC, Van Look KJ, Tyler CR. A critical analysis of the biological impacts of plasticizers on wildlife. Philosophical Transactions of the Royal Society B. 2009;**364**:2047–2062

[102] Lemos MFL, van Gestel, CAM, Soares AMVM. Endocrine disruption in a terrestrial isopod under exposure to bisphenol A and vinclozolin. J Soil Sediments 2009;**9**:492–500.

[103] Kiyomoto M, Kikuchi A, Unuma T, Yokota Y. Effects of ethynylestradiol and bisphenol A on the development of sea urchin embryos and juveniles. Marine Biology. 2006;**149**:57–63

[104] Arslan OC, Parlak H. Effects of bisphenol A on the embryonic development of sea urchin (*Paracentrotus lividus*). Environmental Toxicology. 2008;**23**:387–392

[105] Kontogiannatos D, Swevers L, Zakasis G, Kourti A. The molecular and physiological impact of bisphenol A in *Sesami anonagrioides* (Lepidoptera: Noctuidae). Ecotoxicology. 2015;**24**(2):356–367

[106] Michail X, Kontogiannatos D, Syriou V, Kourti A. Bisphenol-A affects the developmental progression and expression of heat-shock protein genes in the moth *Sesamia nonagrioides*. Ecotoxicology. 2012;**21**(8):2244–2253

[107] Watts MM, Pascoe D. Comparison of *Chironomus riparius Meigen* and *Chironomus tentans Fabricius* (Diptera: Chironomidae) for assessing the toxicity of sediments. Environmental Toxicology and Chemistry. 2000;**19**:1885–1892

[108] Marcial HS, HagiwaraA, Snell TW. Estrogenic compounds affect development of harpacticoid copepod *Tigriopus japonicus*. Environ Toxicol Chem 2003;**22**(12):3025–3030

[109] Watts MM, Pascoe D, Carroll K. Exposure to 17a-ethinylestradiol and bisphenol A e effects on larval moulting and mouthpart structure of *Chironomus riparius*. Ecotoxicol Environ Saf 2003;**54**:207–215

[110] Biggers WJ, Laufer H. Identification of juvenile hormone-active alkylphenols in the lobster *Homarus americanus* and in marine sediments. The Biological Bulletin. 2004;**206**:13–24

[111] Pascoe D, Carroll K, Karntanut W, Watts MM. Toxicity of 17a-ethinylestradiol and bisphenol A to the freshwater cnidarian *Hydra vulgaris*. Arch Environ Contam Toxicol 2002;**43**:56–63

[112] Mansilha C, Silva P, Rocha S, Gameiro P, Domingues V, Pinho C, Ferreira IM. Bisphenol A migration from plastic materials: direct insight of ecotoxicity in Daphnia magna. Environ Sci Pollut Res Int. 2013;**20**(9):6007–18

[113] Tišler T, Krel A, Gerželj U, Erjavec B, Dolenc MS, Pintar A. Environ Pollut. Hazard identification and risk characterization of bisphenols A, F and AF to aquatic organisms.2016;**212**:472–479

[114] Hill M, Stabile C, Steffen LK, Hill A. Toxic effects of endocrine disruptors on freshwater sponges: Common developmental abnormalities. Environmental Pollution. 2002;**117**(2): 295–300

[115] Oehlmann J, Schulte-Oehlmann U, Tillmann M, Markert B. Effects of endocrine disruptors on prosobranch snails (Mollusca: Gastropoda) in the laboratory. Part I: Bisphenol A and octylphenol as xeno-estrogens. Ecotoxicology 2000;**9**:383–397

[116] Oehlmann J, Schulte-Oehlmann U, Bachmann J, Oetken M , Lutz I, Kloas W, Ternes TA. Bisphenol A induces superfeminization in the ramshorn snail Marisa cornuarietis (Gastropoda: Prosobranchia) at environmentally relevant concentrations. Environ Health Perspect 2006;**114**:127–133

[117] Planelló R, Martínez-Guitarte JL, Morcillo G. "http://www.ncbi.nlm.nih.gov/pubmed/18313723"The endocrine disruptor bisphenol A increases the expression of HSP70 and ecdysone receptor genes in the aquatic larvae of Chironomus riparius.Chemosphere 2008;**71**(10):1870–6

[118] Vincent-Hubert F, Revel M, Garric J. DNA strand breaks detected in embryos of the adult snails, Potamopyrgus antipodarum, and in neonates exposed to genotoxic chemicals. Aquatic Toxicology. 2012;**122–123**:1–8

[119] Martínez PP, Morales M, Martínez-Guitarte JS, Morcillo G. Genotoxic effects of environmental endocrine disruptors on the aquatic insect Chironomus riparius evaluated using the comet assay. Mutation Research/Genetic Toxicology and Environmental Mutagenesis 2013;**758**:41–47

[120] Craig ZR, Wang W, Flaws JA. Endocrine-disrupting chemicals in ovarian function: Effects on steroidogenesis, metabolism and nuclear receptor signalling. Reproduction. 2011;**142**(5):633–646

[121] Prossnitz ER, Barton M. The G-protein-coupled estrogen receptor GPER in health and disease. Nature Reviews Endocrinology. 2011;**7**(12):715–726

[122] Weiss B. The intersection of neurotoxicology and endocrine disruption. Neurotoxicology. 2012;**33**(6):1410–1419

[123] Surprenant D. Acute toxicity of bisphenol A to the mysid (*Mysidopsis bahia*) under flow-through conditions. Report BW-85-8-1825, SpringbornBionomics Aquatic Toxicilogy Lab 1985, Wareham, MA, USA

[124] Mihaich EM, Friederich U, Carpers N, Hall AT, Klecka GM, Dimond SS, Staples CA, Ortego LS, Hentges SG. Acute and chronic toxicity testing of bisphenol A with aquatic invertebrates and plants. Ecotoxicology and Environmental Safety. 2009;**72**:1392–1399

Chapter 2

Male Reproduction: One of the Primary Targets of Bisphenol A

Tomáš Jambor, Bistáková Jana, Greifová Hana,
Tvrdá Eva and Lukáč Norbert

Additional information is available at the end of the chapter

http://dx.doi.org/10.5772/intechopen.68629

Abstract

Infertility is a major health issue affecting human life. The most notable factors causing male infertility is exposure to environmental contaminants. Bisphenol A (BPA) is a common toxic environmental contaminant. Human population is exposed to bisphenol A through air, water, food and a variety of industrial products. Growing evidence from research on laboratory animals supports the hypothesis that bisphenol A is able to adversely affect male reproductive function. The specific mechanisms of action of bisphenol A are wide but not definite. Bisphenol A interferes with the hormonal metabolism and regulation, binding affinity or enzymatic activity, resulting in a deviation from a normal reproductive behaviour. Binding ability to androgen and oestrogen receptors, as well as other properties, is currently investigated. A decreased sperm count, inhibition of sperm motility and reduction of organ weights were observed and linked with oxidative stress after bisphenol A treatment. In addition, prenatal exposure to bisphenol A may lead to adverse effects in the offspring. In this review, we address the topic of BPA effects on male reproductive function and emphasize its effects on testicular steroidogenesis and spermatogenesis. A considerably more detailed and systematic research focusing on bisphenol A toxicology is required for a better understanding of risks associated with exposure to this endocrine disruptor.

Keywords: reproduction, male, bisphenol A, steroidogenesis, spermatogenesis

1. Introduction

Over the last decade, research has focused on the potentially hazardous effects of a wide range of chemicals present in the human or wildlife environment. An increased occurrence of male reproductive and developmental disorders such as hypospadias, cryptorchidism and testicular cancer as well as a decreased semen quality have been related to the action of endocrine disruptors. Endocrine-disrupting effects of commercially available products have the potential to cause reproductive dysfunction alongside with adverse effects on development and sexual differentiation. The group of known endocrine disruptors is extremely heterogeneous. One of the most common environmental contaminants classified as endocrine disruptors is bisphenol A (BPA). Many studies have defined BPA as hazardous to the health of humans and animals, particularly to male reproduction [1]. BPA plays a key role in testicular disorders, due to its oestrogenic properties. Oestrogen biosynthesis takes place in the testicular cells; hence the absence of oestrogens causes negative effects on male reproduction [2]. Physiological levels of oestrogens are essential for a normal spermatogenesis; however, overage of oestrogens together with a deficiency of testosterone may cause infertility [3]. In addition, some reports have shown that BPA behaves as an androgen receptor antagonist, interrupting normal androgen receptor binding activity and interactions between androgen receptors and endogenous androgens. Such effects by BPA on the function of endogenous androgens could interfere with normal processes of spermatogenesis, which are controlled by numerous endogenous hormones [4, 5]. Moreover, androgens play characteristic roles in the expression of the male phenotype, development and maintenance of the secondary male characteristics and regulate the expression of an array of target genes that are important for a proper male fertility [6]. As such the chemical substances with antiandrogenic properties can react with male sexual functions and behaviour by blocking the binding of androgens to androgen receptors and a following induced expression of gene by androgen. It has been reported that BPA has adverse effects on the male reproductive system including a decreased sperm count, abnormal sperm motility and reduced reproductive organ weights [7]. One of the potential mechanisms of action of BPA on male reproductive functions and sperm quality has been also proposed to act through oxidative stress. Environmental contaminants such as BPA have been shown to induce reactive oxygen species overgeneration in both intracellular and extracellular spaces leading to cell death and tissue injury [8]. Sensitivity to BPA is not the same at all stages of life, and there are specific critical phases of male development that are more vulnerable to BPA exposure [9]. One such sensitive phase wherein organ differentiation and development take place is the prenatal and perinatal period. The cumulation of BPA in tissues of the male reproductive system is associated with different pathological consequences since low-level BPA exposure during embryonic phase of life has been observed in reduction of effectiveness of spermatogenesis in male descendants [44]. Many experiments have examined the effect of prenatal, neonatal and lactational exposure to low BPA doses. Such studies examined the impact of small dosage of this endocrine disruptor throughout crucial stages of development in various cells and organs. These crucial stages continue throughout reaching of sexual maturity, the physiological phase of modification to fertility [10]. In relation to male reproductive functions, sexual dysfunction in animal studies is difficult to conduct. However, changes in sexual behaviour including a reduced performance in latency and frequency of intromission among rodents exposed to BPA have also been reported [10, 11]. Even results from a human study involving workers of BPA manufactures in China from 2004 to 2008 provide important evidence that occupational exposure to BPA significantly increases the risk of male sexual dysfunction [12]. Another issue that is becoming increasingly debated in the context of male reproductive function and endocrine-disrupting compounds, such as BPA is their ability to modify the epigenome. Hormone cascade pathway is usually returnable and activates constant modulations of cell processes. Throughout sexual development, sex steroids are able to initiate persistent impact on activities of gene that induce developmental changes of cells and genes to react to another hormonal impulse during life. This hormonal imprinting or gene programming probably include mechanisms of epigenetics related to DNA methylation, which can be transmitted from mother cell to daughter cells and cause permanent changes [13]. Multiple evidence from *in vitro* and *in vivo* models have established that epigenetic modifications caused by in utero exposure of BPA can induce alterations in gene expression [14]. Currently, over 2.7 million metric tons of BPA are produced annually, primarily to be used in the production of polycarbonate plastics. This product is a constituent of a wide variety of products, including plastic packaging, water or milk bottles, food wrapping and food cans. BPA can leach into food or beverages from plastic containers and has been found in various human food samples. We may conclude that humans are exposed continuously to BPA primarily through diet [15, 16]. We considered these facts as crucial in the context of a mutual relationship between BPA and potential modifications in male reproductive system.

2. Potential impact of BPA on the steroidogenesis

There is overwhelming evidence about the potential ability of BPA to affect cellular processes, such as steroidogenesis and spermatogenesis. The testicular compartments responsible for steroidogenesis and spermatogenesis are the seminiferous tubules and interstitium. Both are morphologically distinct but functionally connected. Steroidogenesis and spermatogenesis are two vital, high energy demanding processes which are exceptionally vulnerable to damage caused by BPA [17]. Steroidogenesis is a process underway in the Leydig cells. Testosterone as a product of the steroidogenic pathway is released from the Leydig cells under the control of the luteinizing hormone (LH). LH binds to the LH receptor to induce the dissociation of the α subunit of the G protein. Gsα then activates the cyclic adenosine monophosphate (cAMP). cAMP binds to protein kinase A (PKA). The active PKA phosphorylates certain cytoplasmic proteins, which, in turn, will increase the transportation rate of cholesterol into the inner mitochondrial membrane. Cholesterol is then catalyzed by the $P450_{SCC}$ enzyme into pregnenolone. Pregnenolone is delivered to the smooth endoplasmic reticulum and subsequently converted into testosterone. Depending on the species, this conversion can occur via progesterone, 17α-OH progesterone and androstenedione through delta-4 intermediates or via 17α-hydroxypregnenolone, dehydroepiandrosterone and androstenediol as delta-5 intermediates by the actions of enzymes, 3β-hydroxysteroid dehydrogenase (3β-HSD), 17α-hydroxylase and 17β-hydroxysteroid dehydrogenase (17β-HSD) [18]. BPA can alter the level of endogenous steroids at a particular site by altering its synthesis, metabolism, distribution or clearance. Alternatively, the chemicals may interact directly with the steroid receptor to either mimic or block steroid actions [19, 33]. Hormonal activity is mediated by binding to steroid receptors. Specific hormone-receptor complex is translocated to the DNA molecule.

After this step alterations in the expression of steroid—responsive genes—are observed [20]. BPA is able to inhibit the steroidogenic process through specific mechanisms such as binding to the receptors and damage to the steroidogenic enzymes.

2.1. Estrogenic and antiandrogenic affinity of BPA

Although there are different mechanisms through which endocrine disruptors are able to modify the endocrine response, chemical substances that might simulate the effect of steroid hormones by a reaction with their respective receptors continue to receive considerable attention. The initial step in the mechanism of action of steroid hormones is the binding of the steroid to its receptor or binding protein. BPA has been shown to be able to bind to the oestrogen receptor and initiate transcription of the oestrogen receptor—regulated genes *in vitro*. Several studies using laboratory rodent models have found that defects in the reproductive tract were associated with oestrogenic activity of BPA [21, 22, 33]. The effects of BPA on a wide variety of tissues and cells have, until recently, been thought to be mediated by a single nuclear hormone-receptor oestrogen receptor α (ERα). ERα was consistently demonstrated in rodent Leydig cells. In *in vitro* systems, BPA competes with estradiol for binding with the ERα. After the merger of oestrogen with receptors is the complex transported to the specific site of DNA molecule (oestrogen responsive elements), located in the 5′ flanking region. Other components and transcription factors interact with each other and initiate gene transcription [23]. ERα exhibits two distinct gene-transactivating regions, AF1 in the amino terminus and AF2 in the carboxyl terminus [24]. However, the discovery of a second oestrogen receptor β (ERβ) has prompted a re-evaluation of the molecular basis for the oestrogen action. Localization of ERβ has been variable in the mouse Leydig cells. Structurally, ERβ is highly homologous to ERα in the DNA-binding domain (>95% amino acid identity) but shows only a 55% homology in the ligand-binding domain. Furthermore, ERβ shows a discrete tissue distribution being the most predominant oestrogen receptor in the ovary, brain and prostate [25]. According to experimental analysis, BPA can produce major alterations in the context of functional gene product synthesis inside the cells with active ERβ and major coactivator TIF2, although it can be active in the cells with ERα or ERβ and coactivator-1a [26]. However, the validity of these findings has been compromised. BPA is able to induce the expression of the nuclear transcription factor NUR77 in mice Leydig cells, which is involved in LH-mediated testosterone synthesis. After BPA administration, higher activity of protein kinase A and phosphorylation of MAPK mediated with NUR77 expression in Leydig cells was observed. Changes in steroidogenic process within 5 min after administration were detected. This response is too rapid to be mediated by activation of the transcription domains, including classical nuclear oestrogen receptors such as AF1 or AF2. NUR77 mRNA levels were increased above baseline at 1 nm BPA [27]. The oestrogenic activity of BPA was further confirmed by the ability to upregulate the oestrogen target gene expression like pS2 and calbindin-D (9K) in mammalian systems. In fact, BPA is weakly oestrogenic, with a lower potency than the endogenous oestrogen [28, 29]. BPA selectively binds to ERα and ERβ and has a higher affinity for ERβ in the target cells. It was also found that the binding affinity relative to 17β-estradiol for BPA at ERβ was 6.6-fold higher than ERα. Binding of BPA to the oestrogen receptors alters their ability to recruit coactivators that may be important for the differences in tissue-dependent responses [26]. BPA and other alkylphenols stimulated the production of a biomarker of oestrogenic activity, vitellogenin, in male fish. Vitellogenin has been used as a biomarker of exposure to antioestrogenic compounds in numerous *in vivo* and *in vitro* studies using fish. It is induced by an oestrogen-dependent activation of gene expression [30]. According to recent molecular studies, BPA is a selective oestrogen receptor modulator, which means that it acts as an oestrogen agonist in some tissues and an oestrogen antagonist in others. Wersinger et al. [31] demonstrated that male mice with deficient in ERα gene were infertile but had a higher serum testosterone levels than their wild-type siblings, indicating that ERα, albeit along with androgen receptors, has a role in mediating the steroid feedback on the pituitary. Compared to researches on the oestrogenic activities of BPA, data on antiandrogenic activities are controversial. Sohoni and Sumpter [32] reported that BPA exhibits antiandrogenic activities, whilst Gaido et al. [33] emphasized that BPA had no antiandrogenic activity. The chemical substances with antiandrogenic properties are able to react with male reproductive function and behaviour by inhibiting the binding of androgens to androgen receptors (AR) and subsequent androgen-induced gene expression. AR is held as an inactive state, being associated with specific 90 kDA heat-shock proteins before exposure to androgens. Upon ligand binding, androgen receptors are translocated into the nucleus and form a complex with specific DNA sequences called androgen-responsive elements (ARE) to enhance the transcription of target genes recruiting coactivators [34, 35]. In this context, we define coactivators as molecules that interact with nuclear receptors and enhance their transactivation. For example, androgen receptor activator-70 (ARA70) was detected in human prostate cells to enhance the androgen receptor transactivation [36]. So, damages of AR gene at molecular level involve the syndrome of androgen insensitivity (AIS) with specific extent of altered reproductive function development, which is associated with defects of androgen receptor-androgen binding, nuclear import, DNA-binding and transcriptional activation. AIS is an archetypal example of a hormone-resistance disorder. In fact, more mutations have been reported in the AR gene than in any other transcription factors [37]. The impairments associated with action of BPA have motivated some researchers to investigate whether BPA could inhibit androgen receptor binding and subsequent androgen receptor-dependent transcription. Once bound to AR, androgen antagonists are imported into the nucleus excluding endogenous androgens from regulating the androgen-dependent transcription. It has been shown that most of the antiandrogenic chemicals contain at least one aromatic ring with a hydroxyl group. The hydroxyl group on the A-phenyl ring of BPA is essential for the inhibitory effect on the AR transactivation. In addition, the hydrophilic substituent at the methylene bridge of BPA is also an important factor for a higher activity [6]. Lee et al. [38] investigated how AR can be affected. BPA inhibits the interaction of AR and its coactivator, interaction of AR and androgens, nuclear translocation of AR and androgen-induced AR transcriptional activity. The inhibition of androgens following BPA treatment is partial and lacks a dose-response relationship, which suggests that the manner of their inhibition may be noncompetitive. On the other hand, according to Sun et al. [39], BPA has the strongest activity to block the gene expression. When they studied the reason why BPA could inhibit the reporter gene expression, it was found that BPA could compete with 5α-dihydrotestosterone (DHT) to bind to AR. DHT is reduced from testosterone through the action of 5-α-reductase. According to preliminary results, 3,5-substituents of phenol ring of BPA decreased its antiandrogenic activity. Nevertheless 3,5-substituents were reported to increase the oestrogenic and thyroid activity [40]. A lot of substances with oestrogenic properties are implicated in a number of cancers as an initiator, which confirms carcinogenic character. The potential risk is visible in early life stages related to the development of cancer later in life. The proliferation of human prostate cancer was confirmed at low doses (0.1–10 nM) of BPA mediated with mutation of the AR [4]. It was demonstrated that up to 80% of hormone-refractory tumours are characterized by high production of nuclear AR signifying that the receptor is stimulated even without occurrence of competent ligands. Moreover, amplification of the AR has been reported between 22 and 30% of prostate tumours, providing at least one understanding of how the receptor can be stimulated without occurrence of androgens, such as testosterone and testosterone derivates [41]. Though in different experimental animal models exposure of low doses of BPA during embryonal stage of development initiated enlargement of prostate size, enhancement of AR expression, reduced differentiation patterns of the prostate and alterations in secretory activity of the gland [42], on the other side, this evidence is questioned by many several experiments that observed no significant impact of this endocrine-disrupting substance in experimental animals [43]. Based on these conclusions, we may speculate that essential compartments of the reproductive system, such as Leydig cells or spermatozoa, are primary targets of BPA. Some experimental studies have shown that BPA affected spermatogenesis probably by competing with testosterone for the cell binding site or other destructive mechanisms. Degeneration of seminiferous tubules and the loss of elongated spermatids were also demonstrated [44]. Furthermore, there are several studies documenting that BPA not only competes with the oestrogen and testosterone receptor, but it also modifies the gene expression of ERα and ERβ. Furthermore, BPA has been found to induce cell death by inhibiting the testicular endoplasmic reticulum Ca^{2+} pumps [45]. According to recent information, BPA is able to affect steroidogenesis in Leydig cells not only through receptor-mediated response but also to inhibit the steroidogenic enzymatic activity.

2.2. Interaction of BPA and steroidogenic enzymes

There are mechanisms other than ER- or AR-mediated effects through which BPA could affect physiological functions, including modulation of steroidogenesis and interference with metabolic breakdown of oestrogens and detrimental effects on signalling cascades. Examination of the expression of different steroidogenic enzymes provides information on the molecular basis for alterations in hormone biosynthesis caused by exposure to BPA [46–48]. Essential male reproductive hormones are testosterone and androstenedione. Their biosynthesis is called steroidogenesis where steroidogenic enzymes step out as stable components responsible for specific cascades of reactions which transform cholesterol to endogenous male hormones. Steroidogenic processes start with the transport of cholesterol to the mitochondrial inner membrane where the first steroidogenic enzyme cytochrome P450 cholesterol side chain cleavage enzyme (CYP11A1) uses it as a substrate to produce pregnenolone. Pregnenolone subsequently diffuses to the smooth endoplasmic reticulum, where it is converted to testosterone by the enzymes such as 3β-hydroxysteroid dehydrogenase, cytochrome P450 17α-hydroxylase/17,20-lyase (P450c17) and 17β-hydroxysteroid dehydrogenase (17β-HSD). The first reaction in the smooth endoplasmic reticulum is catalyzed by 3β-HSD to progesterone. P450c17 catalyses two reactions that convert progesterone to 17α-hydroxyprogesterone and then to androstenedione. 17β-hydroxysteroid dehydrogenase catalyses the last step from androstenedione to testosterone.

Recent experimental studies have demonstrated that the production of both androstenedione and testosterone was inhibited by BPA in a concentration-dependent manner over the course of 24 h incubation. Lower concentrations of androstenedione and its direct downstream product, testosterone, after the exposure to BPA are consistent with a direct inhibition of enzymatic activities, such as 3β-HSD, cytochrome P450c17 and 17β-HSD. A decrease in the activity of 17α-hydroxylase resulted in a lower production of its direct product 17α-hydroxyprogesterone. Moreover, the decreased activity of 17,20-lyase inhibited the rate of 17α-hydroxyprogesterone conversion to androstenedione, which led to in a 7.7-fold reduction in the androstenedione synthesis and a 2.4-fold reduced testosterone level [49]. It has been reported that prenatal exposure of BPA in rodents causes a reduction in the testosterone production. It is possibly caused by the downregulation of the steroidogenic enzymes in the Leydig cells and an inhibition of LH secretion [50]. Ye et al. [51] confirmed a dose-dependent inhibition of human 3β-HSD and P450c17 by BPA. At 10 μM, BPA also weakly but significantly inhibited human and rat 17β-HSD activities. In general, human steroidogenic enzymes are more sensitive to BPA than rat enzymes. The results also demonstrate that BPA partially competes with cofactor NAD^+ (for 3β-HSD) in the cofactor binding site of this enzyme. The second essential enzyme 17β-HSD, which is responsible for testosterone synthesis from androstenedione, was observed to decrease the activity of this enzyme. 17β-HSD accounts for most of the circulatory testosterones in males and in the case of genetic mutation induced by BPA may cause the autosomal recessive genetic disorder male pseudohermaphroditism in which males often are born with female external genitalia and without a prostate [52, 53]. The aromatase enzyme, which is encoded by the CYP19 gene and catalyses the conversion of androgens to oestrogens, is expressed more in the male reproductive tract than in other tissues in rodents. Some experimental data show that BPA caused a direct inhibition of aromatase gene expression and oestrogen biosynthesis. Disruption of CYP19 gene expression of aromatase in Leydig cells was ERα mediated as oestrogenic agents act via ERα to upregulate the promoter region of the aromatase gene [54, 55]. Some experimental studies indicate that a decreased androgen production by Leydig cells does not correlate with level of testosterone in serum after chronic BPA exposure which can be due to compensatory mechanisms initiated *in vivo*, for example, elevated pituitary luteinizing hormone. Higher serum luteinizing hormone levels in BPA-treated rats presumably resulted from a decreased sensitivity to the androgen negative feedback on the hypothalamus and pituitary and the consequent stimulation of luteinizing hormone secretion [25]. Nikula et al. [56] demonstrated that exposure to environmentally relevant BPA levels had also adverse effects on testicular function by decreasing pituitary LH secretion and reducing Leydig cell steroidogenesis. For example, exposure of pubertal rats to 2.4 μg/kg for 15 days indicated a decreased testosterone levels as well as Leydig cell androgen biosynthetic capacity. On the other hand, small but significant reduction in LH was observed in rats treated with BPA after 2 weeks of treatment, although this effect had disappeared after 5 weeks. Even if BPA inhibits production of testosterone through reduced secretion of LH, it is proved that BPA binds with LH receptor ligand binding, and releasing of LH from the receptor may cause reduction of steroidogenic activity. In the human testes, isoforms of ERβ implying that BPA is able to modulate oestrogen synthesis have been localized. Imbalance in the androgen an oestrogen action, during early stages of differentiation, may induce potential damage of male reproductive parameters and sexual behaviour in adulthood [57]. Both oestrogen and testosterone are necessary for development and functions of male reproductive system tissues; inhibition of steroid hormone biosynthesis can be associated with abnormalities of testes after BPA exposure.

2.3. Effects of BPA exposure during gestation through puberty

We recognize that the development is epigenetic, which refers to changes in gene activity during developments that are mediated by chemical signals. Autocrine, paracrine (growth factors) and endocrine (steroids) signals coordinate the direction of tissue differentiation during critical periods in development. Androgens, mediated by the AR, do play an indispensable role in induction of male sex differentiation and development of the male phenotype. It has been demonstrated that the developing embryo may be much more susceptible to harmful effects of environmental contaminants than adult animals. A high *in vivo* BPA efficiency during embryonic development is caused by low BPA binding to oestrogen-binding proteins in plasma that are intended for control of endogenous estradiol absorption into cells and by the low embryos and newborn capacity to metabolize and inactivate BPA in the liver [58]. Potential BPA cumulation in embryos is supported by findings which demonstrate that BPA status in amniotic fluid at 15–18 weeks of pregnancy is fivefold higher than BPA status in serum of pregnant and non-pregnant females [59]. After treatment of mice during gestation with BPA in dose 100 mg/kg BPA (given subcutaneously), BPA was identified in the brain, liver, foetal sera, uterus and testes 30 min after exposure [60]. Significant effects caused in rats and mice by exposure during development to doses of BPA involve structural and neurochemical changes throughout the brain associated with behavioural changes, such as hyperactivity, learning deficits and increased aggression. Increased aggressiveness in male CD-1 mouse offspring occurred as a result of oral administration of 2 μg/kg/day of BPA to pregnant females [61]. A lot of current researches have observed that BPA treatment of rodents during gestation even in low doses induced persistent impact on tissues of male reproductive organs and female descendants. Recent experiments by Nagel et al. [62] and Vom Saal et al. [63] reported that administration of low oral doses of BPA (2–20 μg/kg/day) to pregnant female mice produced statistically significant increase in the weights of the prostate and decrease in epididymis weights and the efficiency of sperm production in their male offspring. These findings are important for several reasons. Firstly, they provide evidence that microgram volumes of BPA are able to cause teratogenic and genotoxic effects in foetus during gestation. Secondly, they prove that BPA is both absorbed and active after oral administration. When female mice were administered to BPA, in the testicles of BPA-exposed male offspring, expression of anti-Müllerian hormone and steroidogenic acute regulatory protein (StAR) was inhibited, and also size of testicles was reduced. In addition, negative impact was persisting in the sexually developed male offspring at 42 postnatal days [64]. From the viewpoint of organ systems and organisms, sexual steroid hormones are responsible for different implications of male fertility development, such as development and keeping of sexual organ system and secondary sex differences. One of the most interesting findings concerning organizational effects of BPA is decrement in sex differences that are usually evident between sexes. Fascinating is also the effect of this substance only on one of the sexes. The reactions causing diverse impact of BPA within genders are still not fully understood, but there is known mechanism that metabolism of endocrine disruptors, such as BPA, is under the influence of testosterone, whilst BPA is able to alter testosterone metabolism [50, 65]. It follows that particular impact of BPA on male sex in the reproductive organs and tissues is caused by cross-reactions with sex steroid hormones. In pregnant female rats exposed to BPA, serum testosterone levels were decreased in male foetus and pups. The testosterone inhibition is probably induced by BPA-suppressive effect on testicular Leydig cells. In fact, BPA inhibits expression of StAR protein, 17β-HSD and others. Protein expression of luteinizing hormone receptor is also compromised following BPA exposure and may lead to decreasing androgen biosynthesis [66, 67]. Direct mitogenic effect of BPA on the foetal prostate has been demonstrated in some experimental studies. Prostate ductal budding begins in mice 2 days before birth. Prostate development is dependent on DHT production. AR expression in the prostatic mesenchyme is required for directing growth and branching morphogenesis of epithelial buds, presumably by induction of paracrine factors secreted by the mesenchyme [68, 69]. There is evidence that oestrogens modulate the activity of androgens in regulating prostate development. The mesenchyme in mice and rats responds to oestrogens via ERα whereas in the human prostate ERβ. Prostatic growth and androgen receptor ligand-binding activity are permanently decreased in response to high doses of BPA during development [70]. Study of Alonso-Magdalena et al. [71] showed that BPA in small doses throughout sensitive stages of development is able to induce negative effect on glucose homeostasis and insulin sensibility. In this experiment mice exposed to BPA by orally administration were pregnant. These mice exhibited glucose intolerance and increased insulin, glycerol, triglycerides and leptin status in plasma compared to control group of pregnant mice. Currently, there is a lot of evidence that this xenobiotic has severe impact not only on mammals but also on amphibian. *Xenopus laevis* exposed to BPA (10^{-7} mol/L) showed feminization of male sexual characteristics, whereas this impact was significant in female larvae phenotypes compared with control individuals [72]. Primary sexual differentiation in *X. laevis* tadpoles is initiated with an indifferent gonadal phase before differentiation into ovaries or testes occurs. It has been shown that the most sensitive period for the induction of sex reversal in *X. laevis* is between stages 50 and 52 [73]. This sex reversal phenomenon was discovered before, when *X. laevis* tadpoles were treated with 17β-estradiol during gonadal differentiation, and observed that all treated tadpoles developed as fertile females. Exposure to exogenous sex steroids and structurally related substances like BPA, during sensitive phase of sexual differentiation, alters the genetically determined gender of the animals [72, 74]. It is suggested that although the definite fate of the primordial germ cells is determined by genetic factors, alterations in the sex steroid hormonal milieu can override this genetic mechanism. Recently, attention has been drawn to reports of BPA-induced gonadal malformation in either testes or ovaries. In the context of feminization impact, reproductive dysfunction is associated with the development of ovo-testis condition. This phenomenon is characterized as the presence of eggs in the testicular tissue. Ovo-testis structure was observed in some gonochoristic fish species (*Oryzias latipes*) both in laboratory and wild animals exposed to BPA. The induction of ovo-testis was observed in *O. latipes* after 60-day posthatch only in the 1820 μg/L treatment. Growth suppression was also observed in concentration-dependent manner. This suppression might be caused by oestrogenic and alkylphenolic character of BPA [75]. Ashfield et al. [76] suggested that the growth suppression of rainbow trout exposed to alkylphenol chemicals might be influenced by the oestrogenic activity of these chemicals in vitellogenin synthesis, which diverts energy resources from growth. The oestrogenicity of BPA can also prevent anti-Müllerian action on the Müllerian ducts in the male, leading to the feminization of male foetus, and feminization can be initiated via upregulation of genes necessary for normal differentiation of ovary tissues (Foxl2 and Wnt4), with simultaneous inhibition of genes required for testis differentiation (Sox9 and Fgf9) in the foetus [77, 78]. Experimental studies in rodents suggest that BPA causes reproductive toxicity that persists into the second generation. Experimental study of CD-1 mice revealed that exposure to high levels of BPA via ingestion caused a longer gestation period and decreased litter size in the high-dose range. The first female generation appeared to be the most affected as they delivered 51% fewer pups when mated with control partners. The males sired 25% fewer pups in the high BPA group [79].

3. The potential impact of BPA on the spermatogenesis

Spermatogenesis is under the control of the hypothalamic-pituitary-testicular axis and the thyroid gland. Dysfunction of this axis, initiated by endocrine disruptors such as BPA, may result in a discontinuance or alteration of spermatogenesis [80]. BPA acts through sex steroid-mediated hormone cascade pathway to influence functions of reproductive system, and it is likely that BPA is also able to modulate specific characteristics of sexually dimorphic systems, in particular gender differences in the mental functions and behaviours of the sexes [81]. The harmful impact of BPA on male reproductive function may occur over embryonic, pubertal and/or adult life [80]. Many current studies have demonstrated that low doses of this widespread oestrogenic chemical substance can induce strong, membrane-initiated oestrogenic effects [82], indicating that low levels of BPA exposure might interfere with normal oestrogenic signalling pathway [4]. It is known that oestrogen receptors are expressed in the Leydig cells (ERα), whereas ERβ have been described in Sertoli cells, pachytene spermatocytes and round spermatidis of the adult rat and male testis. ER has been also shown to be expressed in other tissues of the male reproductive tract [83]. Recent *in vitro* study has shown that unconjugated (aglycone) BPA binds to ERα and β, generating weak oestrogenic action, and has also high affinity for two membrane-bound oestrogen receptors, G-protein-coupled oestrogen receptor 30 and membrane oestrogen receptor α, in addition to orphan nuclear oestrogen-related receptor γ [84]. Due to previous evidence about biological activity of BPA, in which BPA has the ability to induce division of cultured human breast cancer cells and bind with the ERs, one cannot rule out the possibility that BPA is able to impact process of spermatogenesis in males [83]. Another relevant action by which endocrine disruptors perform their negative impacts on male sexual system is to break the balance between oxidants and antioxidants in testicular tissues, which is associated with the development of oxidative stress and consequent harmful effect on spermatogenesis [85]. The activity of superoxide dismutase, catalase and glutathione peroxidase was reduced, whilst the status of peroxide and peroxidation of lipids significantly increased in the rats exposed to BPA compared to control animals without BPA treatment. Data obtained in this experiment showed that upgrade levels of BPA induce fall of antioxidant defence system and cause oxidative stress in rat epididymal sperm [8]. Also, BPA is administrated during the embryonic/foetal life and over infancy via the placenta and milk; reactive oxygen species were induced in mice testis. Peroxidation of the testis was enhanced and finally resulted in their underdevelopment [86]. *In vitro* studies demonstrated that BPA could cause oxidative stress in sperm cells, which leads to accumulation of free radicals, together with the reduction of cell antioxidant system activity. These oxidative responses were associated with spermatozoa quality decrease, as measured by decline in the rates of spermatozoa motility and velocity. BPA administration also leads to depletion of ATP metabolism and significant DNA fragmentation in sperm cells [87]. Besides that, D'Cruz et al. [88] study suggested that the oestrogenic and oxidative stress-inducing ability of BPA could have supported to the violation of glucose homeostasis in the testis. Results show that sustained exposure to BPA may damage the testicular functions by targeting the glucose metabolism in the testis.

3.1. Spermatogenesis and sperm function affected by BPA

A lot of experiments with BPA have confirmed that this chemical substance, even at levels under doses that are considered as safe for human population, is able to impair sexual functions and behaviour in rodents [89], and for male reproduction, it is proved that the exposure of adult rats to environmental doses of BPA can reduce activity of spermatogenesis and sperm count [8, 83]. *In vivo* study with adult rats suggests that low dosage (2 μg/kg body weight) of BPA after oral administration impairs spermatogenesis by reducing reproductive hormones to stop meiosis of germ cells and activating the Fas/FasL pathway to induce the apoptosis of germ cells, lowers the biosynthesis and secretion of testosterone via inhibiting the activity of GnRH neurons, and decreases the expression of steroidogenic enzymes. Subsequently, declining testosterone rate was accompanied by reduction of sperm concentration [90]. Another recent *in vivo* study with experimental rats evaluated the potential impact of BPA in the doses of 1, 5, 10 and 100 mg/kg body weight on spermatogenesis. Seminiferous tubules were devoid of spermatozoa or were filled with immature germ cells and cellular debris with sloughing of germ cells into the seminiferous tubular lumen. Furthermore, the seminiferous epithelium appeared to be disintegrating with loosening of the intercellular bridges between germ cells as well as between germ cells and sertoli cells. Epididymal tubules also showed empty lumen or lumen filled with cellular debris [91]. Study of Furuya et al. [92] refers to stunted development of testicular tissue in male chickens that were treated for up to 23 weeks by oral administration of BPA in doses low as 2 mg/1000 g body weight every 2 days; chickens receiving BPA showed reduced weight of testes, with the smaller seminiferous tubules exhibiting constrained spermatogenesis. The antispermatogenic potency of BPA proved in experimental animals has been confirmed by several epidemiological studies realized among groups of exposed human males. Epidemiological evidence in humans indicates that urine

BPA levels are highly associated with decline sperm concentrations, indicating an essential association between BPA exposure and production of sperm cells. Compared with men who did not have noticeable urine BPA levels, those with noticeable urine BPA had more than three times the chance of lowered sperm concentration and lower viability of sperm, but there was no found correlation between urine BPA levels and semen volume or sperm morphology [93]. Wang et al. [94] reported an implication between higher urinary BPA concentrations and clinically abnormal thyroid hormones (elevated serum-free T3 levels) that also affect spermatogenesis. Direct action of BPA on spermatozoa is still unclear. Human sperm incubated with 1 μM of BPA showed no significant changes in influxes of calcium and pathological acrosome reaction in the sperm cells. Equally, the genetic material of sperm cells, as evaluated by the TUNEL protocol, comet and redox activity were not damaged by *in vitro* BPA exposure [95]. These facts indicate that the severe BPA impact on male sexual system is caused *in vivo* by different mechanisms, in the concrete alterations in the function of the hypothalamic-pituitary-gonadal axis and thyroid hormone activity [50]. Motility is still the most important parameter in the semen analysis and initial investigation of the male fertility factor, and it is a basic prerequisite that enables sperm cells to fertilize the egg cell [96]. Spermatozoa motility closely relates with mitochondrial activity, because spermatozoa contain many mitochondria helically arranged around the middle piece axonema. Mitochondria play a key role in the energy production through the generation of adenosine triphosphate and maintenance motility of spermatozoa [97]. Possibly, BPA uses endogenous oestrogen signalling pathway and similarly to 17β-estradiol modulates sperm motility. This may be related to its effects on sperm mitochondrial potential and, thus, the generation of ATP. It appears because of the proven middle piece localization of ERs in sperm. This region of sperm cell is characteristic for mitochondrial incidence, and mitochondria are possibly target organelles for oestrogen action [98]. In addition, there is proof that mitochondrial enzymes in the testis such as succinate dehydrogenase, malate dehydrogenase, isocitrate dehydrogenase, monoamine oxidase and NAD dehydrogenase decreased after BPA exposure [99]. Adult male mice show significant reductions in testicular sperm counts, as well as in epididymal sperm counts, after exposure to 25 ng/kg body weight of BPA [100]. Similarly, a reduction of sperm count and motility and an increase of sperm morphological abnormalities, following 2 weeks of BPA administration (10–40 mg/kg body weight), were found [101]. A dose-dependent effect of BPA on male birds was observed also in Singh et al. [102] *in vivo* experiment with physical attributes of semen such as semen volume, sperm motility and sperm concentration. Semen volume was highest in low dose (1 mg/kg body weight), whilst sperm concentration was lowest, indicating tail off of bird semen treated with low-dose BPA. The sperm motility was found smaller in high dose (5 mg/kg body weight) of BPA. It is supposed that high dosage of BPA rises the availability of metabolically active BPA in the blood, which is associated with lower motility parameters of sperm due to blockage of the Ca^{2+} channels activated by voltage or by rising oxidative stress in the sperm cells [103]. The results obtained from *in vitro* study of Lukacova et al. [104] confirm that BPA have the detrimental effect on bovine spermatozoa motility. This study also showed that BPA in different doses (1, 10, 100 and 200 μg/mL) is able to decline mitochondrial activity and spermatozoa viability and caused mitochondrial dysfunction by the increasing intracellular formation of superoxide radical. Analysis of sperm motility parameters confirmed the significant differences between the experimental samples from groups with doses of BPA higher than 100 μg/mL and the control samples. Sperm motility results obtained after 6 h of cultivation at the doses higher than 10 μg/mL of BPA showed significant reduction of motility, and after 24 h cultivation, it was found that the doses less than 10 μg/mL statistically improve motility parameters, whilst the doses higher than 100 μg/mL significantly reduced motility in comparison to samples from control group. Results of motility in this study correlate with other *in vitro* experiments with sperm of various species. Motility parameters of experimental mice sperm were measured following 6h of *in vitro* cultivation with increasing BPA doses (0.0001, 0.01, 1 and 100 μM). Number of motile sperm cells was significantly reduced after incubation with BPA in concentration of 100 μM [105]. Also, another study with chicken sperm showed that environmentally serious concentration of BPA (0.74 mM) significantly decreased motility as well as fertilizing ability, live sperm percentage and mitochondrial membrane potential [102]. Exposure to various BPA concentrations (0.6, 4.5 and 11.0 μg/L) negatively affected motility of fish too [106]. Study with human male observed a demonstrable correlation, although not statistically significant, between BPA exposure, specifically urinary BPA concentration, and altered sperm parameters, such as reduced sperm count affected sperm motility attributes and morphology, and increased damage of DNA integrity in sperm, between infertile men [107]. According to Danish research, 98% of tested men had quantifiable rates of BPA after measuring in urine with an average amount of 3.25 ng/mL. Urine samples from group of tested men with high amounts of BPA also showed considerably higher testosterone, estradiol and luteinizing hormone status than urine samples with lowest amount of BPA. These men showed loss of spermatozoa motility too [108].

BPA experiments on different animals exhibit that impact is usually more damaging throughout in utero stage, which is the most sensitive developmental phase for the foetus. It has been observed that this chemical substance is able to generate different injuries in the foetus, including male embryos feminization, reduced function of the testes and epididymides with breakdown of tissues, enlarged prostate, shortening of anogenital distance and alteration of adult sperm parameters, such as sperm count, motility and density. BPA is also able to affect embryo thyroid development [80]. Recent findings support another additional BPA activity mechanism, by a non-genomic pathway, initiated at membrane receptors, including standard ERs and/or G-protein-coupled receptor 30 [109]. By disrupting levels of hormone or receptor activity, the negative effect of this chemical substance may be to modulate male reproductive organ development throughout foetal life. In addition, harmful BPA impact can be more noticeable and nonreversible throughout this phase of development, in contrast to adults, who reached a functional sex maturity and physiology, in which the harmful impact is eventually not persistent since the first exposure [110]. In utero exposure to BPA was found to cause negative effects on reproductive organs in rodents. In utero exposure of pregnant CD-1 mice to BPA in amount 50 μg BPA/kg body weight/day during 16–18 days of gestation showed increasing the anogenital distance in male young [111]. This conflicts with study of Chahoud et al. [112], who presented shortening of the anogenital distance, following prenatal BPA exposure. However, these studies exhibited that BPA has the ability to alternate anogenital distance during prenatal life. Alteration in the development and tissue organization, changes in prostate gland weight, reduced sperm efficiency and daily sperm production were also observed [58, 62]. Oral administration of 2–20 ng BPA/g body weight to female mice on 11–17 gestational days exhibited significant decline of relative testis weight of male young [61]. Vom Saal et al. [63] researched BPA exposure on male mice during pregnancy and observed raising size for preputial glands and reduced size of epididymides, as well as reduced capacity of daily sperm production. When female mice were co-administered with BPA in combination with di(2-ethylhexyl) phthalate, another chemical plastic substance, the expression level of anti-Müllerian hormone was decreased in the testicular tissue of treated young males and also reduced the size of testes. And more significantly, the negative impact was sustained in the sexually mature young at postnatal day 42, associated with decrease counts of epididymal sperm cells [64]. A decline in fertility, daily sperm production and count and motility of sperm in BPA-exposed male offspring over maturity was also reported in Salian et al. study [113].

3.2. Impact of BPA on Sertoli cell function

Normal function of Sertoli cells that are part of a seminiferous tubule is crucial in the spermatogenesis. Process of differentiation and production of mature sperm cells is under the control of the FSH because Sertoli cells are equipped with FSH receptors on their membranes and are activated by secretion of this adenohypophysis hormone. Inhibition of the Sertoli cell function by BPA, directly or indirectly through reduction of hormone synthesis, may impair reproductive function in exposed males [80]. Sertoli cell function is to provide support, in other words, provide the adequate metabolic and structural background for developing spermatozoa because a lot of factors important for gamete maturation are associated with functions of somatic Sertoli cells. Consequently, any agent that impairs the viability and the function of Sertoli cells may have profound effects on spermatogenesis [114]. Experimental study dealing with impact of BPA on Sertoli cells demonstrates that exposure of cultured Sertoli cells to BPA decreased cell viability. Treated cells showed alterations in morphology, including blebs on membrane, breakdown of cytoskeletal structures, cell rounding and condensation and fragmentation of DNA that conform to the morphological changes of apoptosis. Results strongly suggest that death of BPA-exposed Sertoli cells is not due to necrosis, but to activation of the apoptotic signal pathways in the cells [115]. In cultured Sertoli cells, BPA also has been shown to induce apoptosis. Moreover, BPA-induced damage of Sertoli cells has been reported by blocking endoplasmic reticulum Ca^{2+} homeostasis [116] and the ectoplasmic specialization between Sertoli cells and spermatids [117]. Previous findings suggested this chemical inhibits endoplasmic reticulum Ca^{2+}-ATPase activity and mobilizes intracellular Ca^{2+} concentration in mouse Sertoli cell lines, TM4 [45]. Fiorini et al. [118] also studied mechanism of BPA action on Sertoli cells. Sertoli cells establish intercellular junctions that are essential for spermatogenesis. Currently, it is known that SerW3 Sertoli cells form characteristic protein elements of cell junctions such as gap junctions with connexin 43, tight junctions with occludin and zonula occludens-1 and anchoring junctions with N-cadherin. This xenobiotic substance impairs junctions between adjacent cells in the tissue by decreasing their number or by inducing abnormal position of these membrane proteins within cells. In addition, BPA is also able to induce downregulation of several genes associated with Sertoli cell function (Msi1h, Ncoa1, Nid1, Hspb2 and Gata6) in 6-week-old-male mice after prenatal exposure [119], thereby disrupting the blood-testis barrier and impairing spermatogenesis [120].

3.3. Induction of oxidative stress by BPA in the male reproductive system

Exposure to environmental toxicants such as BPA induces the overproduction of reactive oxygen species, leading to testicular oxidative stress. It is known that BPA decreases the activity of the male-specific cytochrome P450 isoforms, and cytochrome P450 has been shown to induce reactive oxygen species that impairs sperm functions and spermatogenesis [121]. Reactive oxygen species can modify the sperm cytoskeletal and axoneme structures, causing a decrease of sperm motility parameters and low probability of sperm-oocyte fusion and therefore leading to low fertility potential [122]. Free radicals are also able to impair the genetic information within the nucleus of the sperm cell, and this damage to the genome may be translated into infertility [102]. In El-Beshbishy et al. [123] experiment, body weight of BPA orally applied for 14 days to male rats was 10 mg/kg, and considerable decline of enzymes with antioxidant activity in testicular tissue such as catalase, glutathione reductase, superoxide dismutase and glutathione peroxidase has been found. Also, hydrogen peroxide quantity and lipid peroxidation were increased in testes and spermatozoa of BPA-treated animals. Kabuto et al. [86] investigated the modifications in endogenous antioxidant capacity and oxidative damage in the mice testis exposed to BPA, whilst animals were treated with BPA during embryonic and foetal phase of life and during lactation phase by oral administration of drinking water with BPA (5 or 10 μg/L) to their pregnant/lactating mothers and male mice were killed in the fourth week of life. BPA increased levels of thiobarbituric acid-reactive substances in the testis, and results suggested that exposure to this chemical substance induces tissue oxidative stress and peroxidation, ultimately leading to testicular underdevelopment. In another *in vivo* study, testicular antioxidant enzymes were impaired by a very low dose (0.005 mg/kg body weight/day) of BPA following 45 days of exposure [124]. Exposure to BPA may also decline antioxidant enzyme activities and induced lipid peroxidation in both epididymides and sperm cells in rats after administration of BPA (0.2, 2 and 20 μg/kg body weight per day) for 45 days and resulted in inhibition of epididymal motility of sperm and number of sperm depending on the level of dose in treated rats as compared with the corresponding group of control animals [8]. Hulak et al. [87] reported that BPA exposure to fish sperm at concentrations 1.75–10 μg/L is capable of inducing oxidative stress, leading to impaired sperm quality, DNA fragmentation and intracellular adenosine triphosphate content. In research with human spermatozoa, *in vitro* exposure to BPA at concentration starting from 300 μM equally could affect sperm integrity through the induction of prooxidative as well as apoptotic mitochondrial dysfunction. It was associated with an increased mitochondrial generation of superoxide anion, caspase-3 and caspase-9 activation and motility decline [125]. Similar results also provided Lukacova et al. [104] in an *in vitro* study with bovine spermatozoa. Generation of superoxide radical within sperm cells was measured by the NBT assay after 24h incubation with BPA. The results showed that in samples of experimental groups, the quantity of superoxide radical increased unlike to samples in control group without BPA. At the dosages higher than 100 μg/mL of BPA, significant differences were noticed. The viability of spermatozoa in metabolic activity assay (MTT) declines in all experimental groups, but significant differences were observed only at the highest doses of BPA after 24h of *in vitro* cultivation. These results demonstrated that BPA can directly promote biological damage by oxidative stress and induces apoptosis in sperm cells across a range of animal species, including humans. The potential impact of BPA on essential reproductive parameters is presented in **Figure 1**.

3.4. Epigenetic effect of BPA on male reproduction

Some chemicals with oestrogenic properties pass through CYP-mediated redox cycle to quinones. Quinones represent biologically active molecules that can bind by covalent bonds to DNA and proteins occurring in the nucleus, such as DNA and RNA polymerases. *In vitro* study of Atkinson and Roy [126] showed that BPA is oxidized by 70% to bisphenol o-quinone. Authors also postulate that BPA is oxidized first to semiquinone and after that oxidized to bisphenol o-quinone. The chemical reaction of DNA with bisphenol o-quinone produced 6–8 adducts. Quinones and several other reactive components have short half-lives; therefore, negative impact of BPA arises especially in oestrogen-sensitive structures. Given that it is expected that modification to quinones and generating of DNA adducts intervene in structures that binding and retaining BPA. Formation of DNA adducts in tissues of sexual system throughout organogenesis can cause genetical imbalance, gene modifications and chromosomal mutations with permanent effects for mature individuals [127]. Whether irreversible binding of BPS to DNA through metabolic activation may be responsible for some of the toxic effects produced by BPA is not clear. Atkinson and Roy [128] also present an *in vivo* study, where binding BPA to DNA was confirmed. BPA is first converted to hydroxylated BPA. Like catechol BPA then enters into redox reactions. During this redox cycling, BPA is enzymatically oxidized first to semiquinone. Bisphenol semiquinone is then further oxidized to bisphenol-o-quinone. BPA without metabolic activation did not bind to DNA [129]. In con-

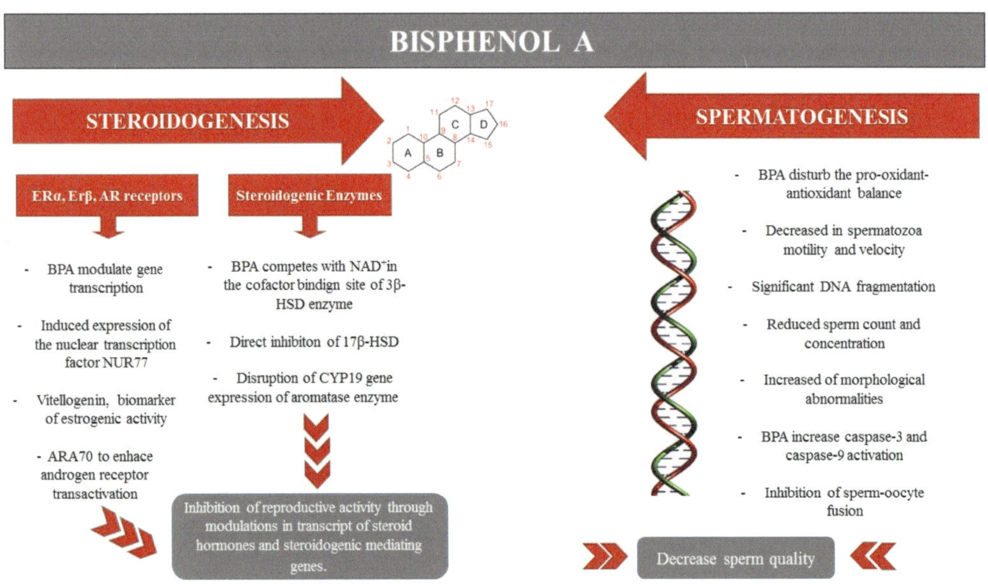

Figure 1. A model summary for the effects of bisphenol A (BPA) on reproductive system. NAD^+, nicotinamide adenine dinucleotide; 3β-HSD, 3β-hydroxysteroid dehydrogenase; 17β-HSD, 17β-hydroxysteroid dehydrogenase.

clusion, directly after metabolic change to metabolites with reactive properties, DNA with covalently converted nucleotides leads to mutational alterations and, therefore, can represent a key attribute in cellular toxicity and development of tumorigenesis process.

Experiments on toxicity of reproductive system exhibited that pregnant female exposed to BPA in prenatal period involved significant fertility disorders of not explicitly F1 male descendants but also subsequent F2 and F3 generations. It also causes increased occurrence of damage during implantation phase in all the three generations. This increase was significant in F3 generation suggesting that this xenobiotic is able to perform its impacts through male germline [130]. Current studies have also begun to suggest the possibility of translation of early exposures to physiological modifications later in life and across generations by epigenetic mechanisms such as methylation-meditated promoter silencing [4]. Epigenetics deals with molecular processes that are associated with hereditary and permanent changes in gene expression. However, these changes do not involve modifications in sequences of DNA. DNA sequences stay constant, but the expression or silencing of genes and regions of gene is mediated by different epigenetic processes, such as methylation, and in reaction to different exposures of environment. DNA methylation is a process by which methyl groups are linked to the DNA molecule, specifically to the cytosine in cytosine-phosphate-guanine segment of DNA. Methylation is able to modify the activity of a DNA sequence without modifying the sequence, and it may cause silencing of gene expression in the segment of DNA [131]. Experiments with rats have demonstrated that BPA exposure and its impact on sexual hormones may cause persistent alterations in the whole male hypothalamic-pituitary-gonadal axis, including development of transgenerational alterations in the levels of steroid hormone receptors in testes, motility of spermatozoa as well as sperm count [113]. Adverse effect of BPA on male germ cells is not matter only prenatal exposure; Tiwari and Vanage [130] experiment demonstrated that adult male rats exposed to 5 mg/kg body weight of BPA during a time of 6 days will generate fatal mutations in spermatozoa. It leads to low sperm motility parameters and sperm production. Due to these facts, it is key to research the epigenetic alterations in male fertility caused by BPA also in later life, not only in critical stages of development. BPA exposure has also been connected to sexually dimorphic alterations in anxiety-like behaviour and general motor activity [132]. This suggests that dose-dependent effects of BPA on emotional aspects and sexually dimorphic manner are associated with demasculinization of characteristic male behaviour [133]. There were also observed alterations in sexual behaviour, especially in a decreased performance in latency and frequency of intromission among BPA-exposed rodents [10, 11]. An impairment in the timing of copulatory sequence was found in Sprague-Dawley male rats, perinatally exposed to BPA via oral administration during pregnancy or lactation [11] or exposed throughout early stages of development [134]. In animals that were postnatally treated by oral administration of BPA, a decrease activity in terms of latence and intromission frequency was noticed [11]. Moreover, same effect in this direction was observed with animals treated in phase of early puberty [134]. Results obtained from study with workers of manufacturers of BPA in China also showed relevant proof that BPA exposure at work significantly increases the possibility of sexual dysfunction in male. The results were the same for all tested parameters that were measured regarding to male sexual dysfunction, all indicating increased risks associated with exposure of BPA. The noticed findings remained after monitoring of wide physiological and psychological aspects that may be related to reproductive disorders between workers exposed to BPA and unexposed workers. Moreover, the relationship between dose and response for found associations also supports the discovery. These findings strengthen probably essential association between exposure to high doses of BPA and raising possibility of reproductive dysfunction in males [93]. Apoptosis of spermatogenic cells was also affected intergenerationally with differential DNA methylation of sperm promoter regions in the F3 generation which was observed in all exposed male lines [135]. Considering the imprinted-like nature of the modified epigenetic DNA methylation sites, sperm cells transfer this epigenome and adult onset disease phenotype to next generations, which is termed epigenetic transgenerational inheritance [136].

Acknowledgements

This study was supported by the Slovak Research and Development Agency Grant nos. APVV-15-0543 and APVV-15-0544.

Author details

Tomáš Jambor, Bistáková Jana, Greifová Hana, Tvrdá Eva and Lukáč Norbert*

*Address all correspondence to: norolukac@gmail.com

Department of Animal Physiology, Faculty of Biotechnology and Food Sciences, Slovak University of Agriculture, Nitra, Slovak Republic

References

[1] Junk GA, Svec HJ, Vick RD, Avery MJ. Contamination of water by synthetic polymer tubes. Environmental Science and Technology. 1974;**8**:1100-1106. DOI: 10.1021/es60098a009

[2] Vitku J, Sosvorova L, Chlupacova T, Hampl R, Hill M, Sobotka V, Heracek J, Bicikova M, Starka L. Differences in bisphenol A and estrogen levels in the plasma and seminal plasma of men with different degrees of infertility. Physiological Research. 2015;**64**:303–311

[3] Pavlovich CP, King P, Goldstein M, Schlegel PN. Evidence of a treatable endocrinopathy in infertile men. The Journal of Urology. 2001;**165**:837–841. DOI: 10.1016/s0022-5347(05)66540-8

[4] Wetherill YB, Akingbemi BT, Kanno J, McLachlan JA, Nadaf A, Sonnenschein C, Watson CS, Zoeller RT, Belcher SM. *In vitro* molecular mechanisms of bisphenol A action. Reproductive Toxicology. 2007;**24**:178–198. DOI: 10.1016/j.reprotox.2007.05.010

[5] Nakamura D, Yanagiba Y, Duan Z, Ito Y, Okamura A, Asaeda N, Tagawa Y, Li C, Taya K, Zhang SY, Naito H, Ramdhan DH, Kamijima M, Nakajima T. Bisphenol A may cause testosterone reduction by adversely affecting both testis and pituitary systems similar to estradiol. Toxicology Letters. 2010;**194**:16–25. DOI: 10.1016/j.toxlet.2010.02.002

[6] Xu LC, Sun H, Chen JF, Bian Q, Qian J, Song L, Wang XR. Evaluation of androgen receptor transcriptional activities of bisphenol A, octylphenol and nonylphenol *in vitro*. Toxicology. 2005;**216**:197–203. DOI: 10.1016/j.tox.2005.08.006

[7] Herath CB, Jin W, Watanabe G, Arai K, Suzuki AK, Taya K. Adverse effects of environmental toxicants, octylphenol and bisphenol A, on male reproductive functions in pubertal rats. Endocrine. 2004;**25**:163–172. DOI: 10.1385/endo:25:2:163

[8] Chitra KC, Latchoumycandane C, Mathur PP. Induction of oxidative stress by bisphenol A in the epididymal sperm of rats. Toxicology. 2003;**185**:119–127. DOI: 10.1016/s0300-483x(02)00597-8

[9] Welshons WV, Nagel SC, Vom Saal FS. Large effects from small exposures. III. Endocrine mechanisms mediating effects of bisphenol A at levels of human exposure. Endocrinology. 2006;**147**:56–69. DOI: 10.1210/en.2005-1159

[10] Richter CA, Birnbaum LS, Farabollini F, Newbold RR, Rubin BS, Talsness CE, Vandenbergh JG, Walser-Kuntz DR, Vom Saal FS. *In vivo* effects of bisphenol A in laboratory rodent studies. Reproductive Toxicology. 2007;**24**:199–224. DOI: 10.1016/j.reprotox.2007.06.004

[11] Farabollini F, Porrini S, Della Seta D, Bianchi F, Dessi-Fulgheri F. Effects of perinatal exposure to bisphenol A on sociosexual behavior of female and male rats. Environmental Health Perspectives. 2002;**110**:409–414. DOI: 10.1289/ehp.02110s3409

[12] Li D, Zhou Z, Qing D, He Y, Wu T, Miao M, Wang J, Weng X, Ferber JR, Herrinton LJ, Zhu Q, Gao E, Checkoway H, Yuan W. Occupational exposure to bisphenol-A (BPA) and the risk of self-reported male sexual dysfunction. Human Reproduction. 2009;**25**:19–27. DOI: 10.1093/humrep/dep381

[13] Kundakovic M, Champagne FA. Epigenetic perspective on the developmental effects of bisphenol A. Brain, Behavior, and Immunity. 2011;**25**:1084–1093. DOI: 10.1016/j.bbi.2011.02.005

[14] Singh S, Li SS. Epigenetic effects of environmental chemicals bisphenol A and phthalates. International Journal of Molecular Sciences. 2012;**13**:10143–10153. DOI: 10.3390/ijms130810143

[15] Shin BS, Kim CH, Jun YS, Kim DH, Lee BM, Yoon CH, Park EH, Lee KC, Han SY, Park KL, Kim HS, Yoo SD. Physiologically based pharmacokinetics of BPA. Journal of Toxicology and Environmental Health A. 2004;**67**:1971–1985. DOI: 10.1080/15287390490514615

[16] Vandenberg LN, Maffini MV, Sonnenschein C, Rubin BS, Soto AM. Bisphenol-A and the great divide: A review of controversies in the field of endocrine disruption. Endocrine Reviews. 2009;**30**:75–95. DOI: 10.1210/er.2008-0021

[17] Huleihel M, Lunenfeld E. Regulation of spermatogenesis by paracrine/autocrine testicular factors. Asian Journal of Andrology. 2004;**6**:259–268

[18] Diemer T, Allen JA, Hales KH, Hales DB. Reactive oxygen disrupts mitochondria in MA-10 tumor Leydig cells and inhibits steroidogenic acute regulatory (StAR) protein and steroidogenesis. Endocrinology. 2003;**144**:2882–2891. DOI: 10.1210/en.2002-0090

[19] Labohá P, Jambor T, Yawer A, Lukáč N, Sovadinová I. Molecular mechanisms of alkylphenol-mediated endocrine disruption in Leydig cells. Toxicology Letters. 2016;**258**:245–246. DOI: 10.1016/j.toxlet.2016.06.1872

[20] Tsai M, O'Malley BW. Molecular mechanisms of action of steroid/thyroid receptor superfamily members. Annual Review of Biochemistry. 1994;**63**:451–486. DOI: 10.1146/annurev.biochem.63.1.451

[21] Danzo BJ. Environmental xenobiotics may disrupt normal endocrine function by interfering with the binding of physiological ligand to steroid receptors and binding proteins. Environmental Health Perspectives. 1997;**105**:294–301. DOI: 10.2307/3433266

[22] Bolger R, Wiese TE, Ervin K, Nestich S, Checovich W. Rapid screening of environmental chemicals for estrogen receptor binding capacity. Environmental Health Perspectives. 1998;**106**:551–557. DOI: 10.2307/3434229

[23] Gould JC, Leonar LS, Maness SC, Wagner BL, Conner K, Zacharewski T, Safe S, McDonnell DP, Gaido KW. Bisphenol A interacts with the estrogen receptor alpha in a distinct manner from estradiol. Molecular and Cellular Endocrinology. 1998;**142**:203–214. DOI: 10.1016/s0303-7207(98)00084-7

[24] Tora L, White J, Brou Ch, Tasset D, Webster N, Scheer E, Chambon P. The human estrogen receptor has two independent nonacidic transcriptional activation functions. Cell. 1989;**59**:477–487. DOI: 10.1016/0092-8674(89)90031-7

[25] Pennie WD, Aldridge TC, Brooks AN. Differential activation by xenoestrogens of ERα and ERβ when linked to different response elements. Journal of Endocrinology. 1998;**158**:11–14. DOI: 10.1677/joe.0.158r011

[26] Routledge EJ, White R, Parker MG, Sumpter JP. Differential effects of xenoestrogens on coactivator recruitment by estrogen receptor (ER) alpha and ER beta. The Journal of Biological Chemistry. 2000;**275**:35986–35993. DOI: 10.1074/jbc.m006777200

[27] Song KH, Lee K, Choi HS. Endocrine disrupter bisphenol A induces orphan nuclear receptor Nur77 gene expression and steroidogenesis in mouse testicular Leydig cells. Endocrinology. 2002;**143**:2208–2215. DOI: 10.1210/endo.143.6.8847

[28] An BS, Kang SK, Shin JH, Jeung EB. Stimulation of calbindin-D(9k) mRNA expression in the rat uterus by octylphenol, nonylphenol and bisphenol. Molecular and Cellular Endocrinology. 2002;**191**:177–186. DOI: 10.1016/s0303-7207(02)00042-4

[29] Vivacqua A, Recchia AG, Fasanella G, Gabriele S, Carpino A, Rago V, Di Gioia ML, Leggio A, Bonofiglio D, Liguori A, Maggiolini M. The food contaminants bisphenol A

and 4-nonylphenol act as agonists for estrogen receptor alpha in MCF7 breast cancer cells. Endocrine. 2003;**22**:275–284. DOI: 10.1385/endo:22:3:275

[30] Rankouhi R, Sanderson JT, Van Holstejin I, Van Kooten P, Bosveld AT, Van den Berg M. Effects of environmental and natural estrogens on vitellogenin production in hepatocytes of the brown frog (*Rana temporaria*). Aquatic Toxicology. 2005;**71**:97–101. DOI: 10.1016/j.aquatox.2004.09.009

[31] Wersinger SR, Haisenleder DJ, Lubahn DB, Rissman EF. Steroid feedback on gonadotropin release and pituitary gonadotropin subunit mRNA in mice lacking a functional estrogen receptor alpha. Endocrine. 1999;**11**:137–143. DOI: 10.1385/endo:11:2:137

[32] Sohoni P, Sumpter JP. Several environmental oestrogens are also anti-androgens. Journal of Endocrinology. 1998;**158**:327–339. DOI: 10.1677/joe.0.1580327

[33] Gaido KW, Maness SC, McDnnell DP, Dehal SS, Kupfer D, Safe S. Interaction of methoxychlor and related compounds with estrogen receptor alpha and beta, and androgen receptor: Structure-activity studies. Molecular Pharmacology. 2000;**58**:852–858. DOI: 10.1124/mol.58.4.852

[34] Wong CH, Zhou ZX, Sar M, Wilson EM. Steroid requirement for androgen receptor dimerization and DNA binding. The Journal of Biological Chemistry. 1993;**268**:19004–19012

[35] McKenna NJ, Lanz RB, O'Malley BW. Nuclear receptor coregulators: Cellular and molecular biology. Endocrine Reviews. 1999;**20**:321–344. DOI: 10.1210/er.20.3.321

[36] Yeh S, Chang C. Cloning and characterization of a specific coactivator, ARA70, for the androgen receptor in human prostate cells. Proceedings of the National Academy of Sciences. 1996;**93**:5517–5521. DOI: 10.1073/pnas.93.11.5517

[37] Quigley CA, De Bellis A, Marschke KB, el-Awady MK, Wilson EM, French FS. Androgen receptor defects: Historical, clinical, and molecular perspectives. Endocrine Reviews. 1995;**16**:271–321. DOI: 10.1210/er.16.3.271

[38] Lee HJ, Chattopadhyay S, Gong EY, Ahn RS, Lee K. Antiandrogenic effects of bisphenol A and nonylphenol on the function of androgen receptor. Toxicological Sciences. 2003;**75**:40–46. DOI: 10.1093/toxsci/kfg150

[39] Sun H, Xu LC, Chen JF, Song L, Wang XR. Effect of bisphenol A, tetrachlorobisphenol A and pentachlorophenol on the transcriptional activities of androgen receptor-mediated reporter gene. Food and Chemical Toxicology. 2006;**44**:1916–1921. DOI: 10.1016/j.fct.2006.06.013

[40] Kitamura S, Suzuki T, Sanoh S, Kohta R, Jinno N, Suigihara K, Yoshihara S, Fujimota N, Watanabe H, Ohta S. Comparative study of the endocrine-disrupting activity of bisphenol A and 19 related compounds. Toxicological Sciences. 2005;**84**:249–259. DOI: 10.1093/toxsci/kfi074

[41] Visakorpi T, Hyytinen E, Koivisto P, Tanner M, Keinänen R, Palmgerg Ch, Palotie A, Tammela T, Isola J, Kallioniemi OP. *In vivo* amplification of the androgen receptor

gene and progression of human prostate cancer. Nature Genetics. 1995;**9**:401–406. DOI: 10.1038/ng0495-401

[42] Ramos JG, Varayoud J, Sonnenschein C, Soto AM, Toro MM, Luque EH. Prenatal exposure to low doses of bisphenol A alters the periductal stroma and glandular cell function in the rat ventral prostate. Biology of Reproduction. 2001;**65**:1271–1277. DOI: 10.1095/biolreprod65.4.1271

[43] Cagen SZ, Waechter JM, Dimond SS, Breslin WJ, Butala JH, Jekat FW, Joiner RL, Shiotsuka RN, Veenstra GE, Harris LR. Normal reproductive organ development in CF-1 mice following prenatal exposure to bisphenol A. Toxicological Sciences. 1999;**50**:36–44. DOI: 10.1093/toxsci/50.1.36

[44] Takahashi O, Oishi S. Disposition of orally administered 2,2-bis (4-hydroxyphenyl) propane (BPA) in pregnant rats and the placental transfer to fetuses. Environmental Health Perspectives. 2000;**108**:931–935. DOI: 10.2307/3435050

[45] Hughes PJ, McLellan H, Lowes DA, Kahn SZ, Bilmen JG, Tovey SC, Godfrey RE, Michell RH, Kirk CJ, Michelangeli F. Estrogenic alkylphenols induce cell death by inhibiting testis endoplasmic reticulum $CA(2^{+})$ pumps. Biochemical and Biophysical Research Communications. 2000;**277**:568–574. DOI: 10.1006/bbrc.2000.3710

[46] Sanderson JT, Seinen W, Giesy JP, van den Berg M. 2-chloro-s-triazine herbicides induce aromatase (CYP19) activity in H295R human adrenocortical carcinoma cells: A novel mechanism for estrogenicity. Toxicological Sciences. 2000;**54**:121–127. DOI: 10.1093/toxsci/54.1.121

[47] Hilscherova K, Jones PD, Gracia T, Newsted JL, Zhang XW, Sanderson JT, Yu RMK, Wu RSS, Giesy JP. Assessment of the effects of chemicals on the expression of ten steroidogenic genes in the H295R cell line using real-time PCR. Toxicological Sciences. 2004;**81**:78–89. DOI: 10.1093/toxsci/kfh191

[48] Zhang X, Yu RM, Jones PD, Lam GK, Newsted JL, Gracia T, Hecker M, Hilscherova K, Sanderson T, Wu RS, Giesy JP. Quantitative RT-PCR methods for evaluating toxicant-induced effects on steroidogenesis using the H295R cell line. Environmental Science and Technology. 2005;**39**:2777–2785. DOI: 10.1021/es048679k

[49] Zhang X, Chang H, Wiseman S, He Y, Higley E, Jones P, Wong CK, Al-Khedhairy A, Giesy JP, Hecker M. Bisphenol A disrupts steroidogenesis in human H295R cells. Toxicological Sciences. 2011;**121**:320–327. DOI: 10.1093/toxsci/kfr061

[50] Akingbemi BT, Sottas CHM, Koulova AI, Klinefelter GR, Hardy MP. Inhibition of testicular steroidogenesis by the xenoestrogen bisphenol A is associated with reduced pituitary luteinizing hormone secretion and decreased steroidogenic enzyme gene expression in rat Leydig cells. Endocrinology. 2004;**145**:592–603. DOI: 10.1210/en.2003-1174

[51] Ye X, Wong LY, Bishop AM, Calafat AM. Variability of urinary concentrations of bisphenol A in spot samples, first-morning voids, and 24-hour collections. Environmental Health Perspectives. 2011;**119**:983–988. DOI: 10.1289/ehp.1002701

[52] Geissler WM, Davis DL, Wu L, Bradshaw KD, Patel S, Mendonca BB, Elliston KO, Wilson JD, Russell DW, Andersson S. Male pseudohermaphroditism caused by mutations of testicular 17 beta-hydroxysteroid dehydrogenase 3. Nature Genetics. 1994;**7**:34–39. DOI: 10.1038/ng0594-34

[53] Andersson S, Moghrabi N. Physiology and molecular genetics of 17 beta-hydroxysteroid dehydrogenases. Steroids. 1997;**62**:143–147. DOI: 10.1016/s0039-128x(96)00173-0

[54] Hess RA. Oestrogen in fluid transport in efferent ducts of the male reproductive tract. Reviews of Reproduction. 2000;**5**:84–92. DOI: 10.1530/revreprod/5.2.84

[55] Kinoshita Y, Chen S. Induction of aromatase (CYO19) expression in breast cancer cells through a nongenomic action of estrogen receptor α. Cancer Research. 2003;**63**:3546–3555

[56] Nikula H, Talonpoika T, Kaleva M, Toppari J. Inhibition of hCG-stimulated steroidogenesis in cultured mouse Leydig tumor cells by BPA and octylphenols. Toxicology and Applied Pharmacology. 1999;**157**:166–173. DOI: 10.1006/taap.1999.8674

[57] Gaskell TL, Robinson LL, Groome, NP, Anderson RA, Saunders PT. Differential expression of two estrogen receptor-β isoforms in the human fetal testis during the second trimester of pregnancy. The Journal of Clinical Endocrinology and Metabolism. 2003;**88**:424–432. DOI: 10.1210/jc.2002-020811

[58] Schönfelder G, Wittfoht W, Hopp H, Talsness CHE, Paul M, Chahoud I. Parental bisphenol A accumulation in the human maternal-fetal-placental unit. Environmental Health Perspectives. 2002;**110**:703–707. DOI: 10.1289/ehp.021100703

[59] Ikezuki Y, Tsutsumi O, Takai Y, Kamei Y, Taketani Y. Determination of bisphenol A concentrations in human biological fluids reveals significant early prenatal exposure. Human Reproduction. 2002;**17**:2839–2841. DOI: 10.1093/humrep/17.11.2839

[60] Uchida K, Suzuki A, Kobayashi Y, Buchanan DL, Sato T, Watanabe H, Katsu Y, Suzuki J, Asaoka K, Mori Ch, Arizono K, Iguchi T. Bisphenol A administration during pregnancy results in fetal exposure in mice and monkeys. Journal of Health Science. 2002;**48**:579–582. DOI: 10.1248/jhs.48.579

[61] Kawai K, Nozaki T, Nishikata H, Aou S, Takii M, Kubo Ch. Aggressive behavior and serum testosterone concentration during the maturation process of male mice: The effects of fetal exposure to bisphenol A. Environmental Health Perspectives. 2003;**111**:175–178. DOI: 10.1289/ehp.5440

[62] Nagel SC, Vom Saal FS, Thayer KA, Dhar MG, Boechler M, Welshons WV. Relative binding affinity-serum modified access (RBA-SMA) assay predicts the relative *in vivo* bioactivity of the xenoestrogens bisphenol A and octylphenol. Environmental Health Perspectives. 1997;**105**:70–76. DOI: 10.2307/3433065

[63] Vom Saal FS, Cooke PS, Buchanan DL, Palanza P, Thayer KA, Nagel SC, Parmigiani S, Welshons WV. A physiologically based approach to the study of bisphenol A and other estrogenic chemicals on the size of reproductive organs, daily sperm production, and behavior. Toxicology and Industrial Health. 1998;**14**:239–260. DOI: 10.1177/074823379801400115

[64] Xi W, Wan HT, Zhao YG, Wong MH, Giesy JP, Wong CK. Effects of perinatal exposure to bisphenol A and di(2-ethylhexyl)-phthalate on gonadal development of male mice. Environmental Science and Pollution Research International. 2012;**19**:2515–2527. DOI: 10.1007/s11356-012-0827-y

[65] Takeuchi T, Tsutsumi O. Serum bisphenol A concentrations showed gender differences, possibly linked to androgen levels. Biochemical and Biophysical Research Communications. 2002;**291**:76–78. DOI: 10.1006/bbrc.2002.6407

[66] Tanaka M, Nakaya S, Katayama M, Leffers H, Nozawa S, Nakazawa R, Iwamoto T, Kobayashi S. Effect of prenatal exposure to bisphenol A on the serum testosterone concentration of rats at birth. Human and Experimental Toxicology. 2006;**25**:369–373. DOI: 10.1191/0960327106ht638oa

[67] Nanjappa MK, Simon L, Akingbemi BT. The industrial chemical bisphenol A (BPA) interferes with proliferative activity and development of steroidogenic capacity in rat Leydig cells. Biology of Reproduction. 2012;**86**:1–12. DOI: 10.1095/biolreprod.111.095349

[68] Timms BG, Petersen SL, Vom Saal FS. Prostate gland growth during development is stimulated in both male and female rat fetuses by intrauterine proximity to female fetuses. The Journal of Urology. 1999;**161**:1694–1701. DOI: 10.1016/s0022-5347(05)69007-6

[69] Marker PC, Donjacour AA, Dahiya R, Cunha GR. Hormonal, cellular, and molecular control of prostatic development. Developmental Biology. 2003;**253**:165–174. DOI: 10.1016/s0012-1606(02)00031-3

[70] Adam BL, Qu Y, Davis JW, Ward MD, Clements MA, Cazares LH, Semmes OJ, Schellhammer PF, Yasui Y, Feng Z, Wright GL. Serum protein fingerprinting coupled with a pattern-matching algorithm distinguishes prostate cancer from benign prostate hyperplasia and healthy men. Cancer Research. 2002;**62**:3609–3614

[71] Alonso-Magdalena P, Vieira E, Soriano S, Menes L, Burks, D, Quesada I, Nadal A. Bisphenol A exposure during pregnancy disrupts glucose homeostasis in mother and adult male offspring. Environmental Health Perspectives. 2010;**118**:1243–1250. DOI: 10.1289/ehp.1001993

[72] Kloas W, Lutz I, Einspanier R. Amphibians as a model to study endocrine disruptors: II. Estrogenic activity of environmental chemicals *in vitro* and *in vivo*. The Science of the Total Environment. 1999;**225**:59–68. DOI: 10.1016/s0048-9697(98)00332-5

[73] Miyata S, Koike S, Kubo T. Hormonal reversal and the genetic control of sex differentiation in *Xenopus*. Zoological Science. 1999;**16**:335–340. DOI: 10.2108/zsj.16.335

[74] Bögi C, Levy G, Luts I, Kloas W. Functional genomics and sexual differentiation in amphibians. Comparative biochemistry and Physiology. Part B , Biochemistry & Molecular Biology. 2002;**133**:559–570. DOI: 10.1016/s1096-4959(02)00162-8

[75] Yokota H, Tsuruda Y, Maeda M, Oshima Y, Tadokoro H, Nakazono A, Honjo T, Kobayashi K. Effect of bisphenol A on the early life stage in Japanese medaka (*Oryzias latipes*). Environmental Toxicology and Chemistry. 2000;**19**:1925–1930. DOI: 10.1002/etc.5620190730

[76] Ashfield LA, Pottinger TG, Sumpter JP. Exposure of female juvenile rainbow trout to alkylphenolic compounds results in modifications to growth and ovosomatic index. Environmental Toxicology. 1998;**17**:679–686. DOI: 10.1897/1551-5028(1998)017<0679:eof jrt>2.3.co;2

[77] Pryor JL, Hughes C, Foster W, Hales BF, Rodaire B. Critical windows of exposure for children's health: The reproductive system in animals and humans. Environmental Health Perspectives. 2000;**108**:491–503. DOI: 10.2307/3454541

[78] Aoki T, Takada T. Bisphenol A modulates germ cell differentiation and retinoic acid signaling in mouse ES cells. Reproductive Toxicology. 2012;**34**:463–470. DOI: 10.1016/j.reprotox.2012.06.001

[79] Morrissey RE, Lamb JC, Schwetz BA, Teague JL, Morris RW. Association of sperm, vaginal cytology, and reproductive organ weight data with results of continuous breeding reproduction studies in Swiss (CD-1) mice. Fundamental and Applied Toxicology. 1988;**11**:359–371. DOI: 10.1016/0272-0590(88)90160-1

[80] Manfo FPT, Jubendradass R, Nantia EA, Moundipa PF, Mathur PP. Adverse effects of bisphenol A on male reproductive function. Reviews of Environmental Contamination and Toxicology. 2014;**228**:57–82. DOI: 10.1007/978-3-319-01619-1_3

[81] Kato H, Furuhashi T, Tanaka M, Katsu Y, Watanabe H, Ohta Y, Iguchi T. Effects of BPA given neonatally on reproductive functions of male rats. Reproductive Toxicology. 2006;**22**:20–29. DOI: 10.1016/j.reprotox.2005.10.003

[82] Wozniak AL, Bulayeva NN, Watson CS. Xenoestrogens at picomolar to nanomolar concentrations trigger membrane estrogen receptor-α-mediated Ca^{2+} fluxes and prolactin release in GH3/B6 pituitary tumor cells. Environmental Health Perspectives. 2005;**113**:431–439. DOI: 10.1289/ehp.7505

[83] Sakaue M, Ohsako S, Ishimura R, Kurosawa S, Kurohmaru M, Hayashi Y, Aoki Y, Yonemoto J, Tohyama C. Bisphenol-A affects spermatogenesis in the adult rat even at low dose. Journal of Occupational Health. 2001;**43**:185–190. DOI: 10.1539/joh.43.185

[84] Mínquez-Alarcón L, Hauser R, Gaskins AJ. Effects of bisphenol A on male and couple reproductive health: A review. Fertility and Sterility. 2016;**106**:864–870. DOI: 10.1016/j.fertnstert.2016.07.1118

[85] Mathur PP, D'Cruz SC. The effect of environmental contaminants on testicular function. Asian Journal of Andrology. 2011;**13**:585–591. DOI: 10.1038/aja.2011.40

[86] Kabuto H, Amakawa M, Shishibori T. Exposure to bisphenol A during embryonic/fetal life and infancy increases oxidative injury and causes underdevelopment of the brain and testis in mice. Life Sciences. 2004;**74**:2931–2940. DOI: 10.1016/j.lfs.2003.07.060

[87] Hulak M, Gazo I, Shaliutina A, Linhartova P. *In vitro* effects of BPA on the quality parameters, oxidative stress, DNA integrity and adenosine triphosphate content in sterlet (*Acipenser ruthenus*) spermatozoa. Comparative Biochemistry and Physiology. 2013;**158**:64–71. DOI: 10.1016/j.cbpc.2013.05.002

[88] D'Cruz SC, Jubendradass R, Mathur PP. Bisphenol A induces oxidative stress and decreases levels of insulin receptor substrate 2 and glucose transporter 8 in rat testis. Reproductive Sciences. 2012;**19**:163–172. DOI: 10.1177/1933719111415547

[89] Rubin BS, Murray MK, Damassa DA, King JC, Soto AM. Perinatal exposure to low doses of bisphenol A affects body weight, patterns of estrous cyclicity, and plasma LH levels. Environmental Health Perspectives. 2001;**109**:675–680. DOI: 10.2307/3454783

[90] Jin P, Wang X, Chang F, Bai Y, Li Y, Zhou R, Chen L. Low dose bisphenol A impairs spermatogenesis by suppressing reproductive hormone production and promoting germ cell apoptosis in adult rats. Journal of Biomedical Research. 2013;**27**:135–144. DOI: 10.7555/jbr.27.20120076

[91] Gurmeet KSS, Rosnah I, Normadiah MK, Das S, Mustafa AM. Detrimental effects of BPA on development and functions of the male reproductive system in experimental rats. Experimental and Clinical Sciences Journal 2014;**13**:151–160

[92] Furuya M, Adachi K, Kuwahara S, Ogawa K, Tsukamoto Y. Inhibition of male chick phenotypes and spermatogenesis by bisphenol-A. Life Sciences. 2006;**78**:1767–1776. DOI: 10.1016/j.lfs.2005.08.016

[93] Li D, Zhou Z, Miao M, He Y, Wang J, Ferber J, Herrinton LJ, Gao E, Yuan W. Urine bisphenol-A (BPA) level in relation to semen quality. Fertility and Sterility. 2011;**95**:625–630. DOI: 10.1016/j.fertnstert.2010.09.026

[94] Wang T, Lu J, Xu M, Xu Y, Li M, Liu Y, Tian X, Chen Y, Dai M, Wang W, Lai S, Bi Y, Ning G. Urinary bisphenol A concentration and thyroid function in Chinese adults. Epidemiology. 2013;**24**:295–302. DOI: 10.1097/ede.0b013e318280e02f

[95] Bennetts LE, De Iuliis GN, Nixon B, Kime M, Zelski K, McVicar CM, Lewis SE, Aitken RJ. Impact of estrogenic compounds on DNA integrity in human spermatozoa: Evidence for cross-linking and redox cycling activities. Mutation Research/Fundamental and Molecular Mechanisms of Mutagenesis. 2008;**641**:1–11. DOI: 10.1016/j.mrfmmm.2008.02.002

[96] Martinez C, Mar C, Azcarate M, Pascual P, Aritzeta JM, Lopez-Urrutia A. Sperm motility index: A quick screening parameter from sperm quality analyser-IIB to rule out oligo- and asthenozoospermia in male fertility study. Human Reproduction. 2000;**15**:1727–1733. DOI: 10.1093/humrep/15.8.1727

[97] Jansen RPS, Burton GJ. Mitochondrial dysfunction in reproduction. Mitochondrion. 2004;**4**:577–600. DOI: 10.1016/j.mito.2004.07.038

[98] Kotwicka M, Skibinska I, Jendraszak M, Jedrzejczak P. 17β-estradiol modifies human spermatozoa mitochondrial function in vitro. Reproductive Biology and Endocrinology. 2016;**14**:50. DOI: 10.1186/s12958-016-0186-5

[99] Anjum S, Rahman A, Kaur M, Ahmad F, Rashid H, Ansari RA, Raisuddin S. Melatonin Ameliorates BPA-induced biochemical toxicity in testicular mitochondria of mouse. Food and Chemical Toxicology. 2011;**49**:2849–2854. DOI: 10.1016/j.fct.2011.07.062

[100] Al-Hiyasat AS, Darmani H, Elbetieha AM. Effects of bisphenol A on adult male mouse fertility. European Journal of Oral Sciences. 2002;**110**:163–167. DOI: 10.1034/j.1600-0722.2002.11201.x

[101] Dobrzyńska MM, Radzikowska J. Genotoxicity and reproductive toxicity of bisphenol A and X-ray/bisphenol A combination in male mice. Drug and Chemical Toxicology. 2013;**36**:19–26. DOI: 10.3109/01480545.2011.644561

[102] Singh RP, Shafeeque CM, Sharma SK, Pandey NK, Singh R, Mohan J, Kolluri G, Saxena M, Sharma B, Sastry KV, Kataria JM, Azeez PA. Bisphenol A reduces fertilizing ability and motility by compromising mitochondrial function of sperm. Enviromental Toxicology and Chemistry. 2015;**34**:1617–1622. DOI: 10.1002/etc.2957

[103] Deutschmann A, Hans M, Meyer R, Häberlein H, Swandulla D. Bisphenol A inhibits voltage-activated Ca^{2+} channels *in vitro*: Mechanisms and structural requirements. Molecular Pharmacology. 2013;**83**:501–511. DOI: 10.1124/mol.112.081372

[104] Lukacova J, Jambor T, Knazicka Z, Tvrda E, Kolesarova A, Lukac N. Dose- and time-dependent effects of BPA on bovine spermatozoa *in vitro*. Journal of Environmental Science and Health. 2015;**50**:669–676. DOI: 10.1080/10934529.2015.1011963

[105] Rahman MS, Kwon WS, Lee JS, Yoon SJ, Ryu BY, Pang MG. Bisphenol-A affects male fertility via fertility-related proteins in spermatozoa. Scientific Reports. 2015;**5**:9169. DOI: 10.1038/srep09169

[106] Hatef A, Alavi SMH, Linhartova Z, Rodina M, Policar T, Linhart O. *In vitro* effects of BPA on sperm motility characteristics perca fluviatilis L. (Percidae; Teleostei). Journal of Applied Ichthyology. 2010;**26**:696–701. DOI: 10.1111/j.1439-0426.2010.01543.x

[107] Meeker JD, Ehrlich S, Toth TL, Wright DL, Calafat AM, Trisini AT, Ye X, Hauser R. Semen quality and sperm DNA damage in relation to urinary bisphenol A among men from an infertility clinic. Reproductive Toxicology. 2010;**30**:532–539. DOI: 10.1016/j.reprotox.2010.07.005

[108] Lassen TH, Frederiksen H, Jensen TK, Petersen JH, Joensen UN, Main KM, Skakkebaek NE, Juul A, Jorgensen N, Andersson A. Urinary bisphenol A levels in young men: Association with reproductive hormones and semen quality. Environmental Health Perspectives. 2014;**122**:478–484. DOI: 10.1289/ehp.1307309

[109] Iwakura T, Iwafuchi M, Muraoka D, Yokosuka M, Shiga T, Watanabe C, Ohtani-Kaneko R. *In vitro* effects of bisphenol A on developing hypothalamic neurons. Toxicology. 2010;**272**:52–58. DOI: 10.1016/j.tox.2010.04.005

[110] Miao M, Yuan W, He Y, Zhou Z, Wang J, Gao E, Li G, Li D. In utero exposure to bisphenol-A and anogenital distance of male offspring. Birth Defects Research. 2011;**91**:867–872. DOI: 10.1002/bdra.22845

[111] Gupta CH. Reproductive malformation of the male offspring following maternal exposure to estrogenic chemicals. Proceedings of the Society for Experimental Biology and Medicine. 2000;**224**:61–68. DOI: 10.1046/j.1525-1373.2000.22402.x

[112] Chahoud I, Fialkowski O, Gericke CH, Merker H, Talsness CE. The effects of low and high doses of bisphenol A on the reproductive system of female and male rat offspring. Reproductive Toxicology. 2000;**40**:587–599. DOI: 10.1016/S0890-6238(01)00153-8

[113] Salian S, Doshi T, Vanage G. Perinatal exposure of rats to bisphenol A affects fertility of male offspring-an overview. Reproductive Toxicology. 2011;**31**:359–362. DOI: 10.1016/j.reprotox.2010.10.008

[114] Griswold MD. Interactions between germ cells and Sertoli cells in the testis. Biology of Reproduction. 1995;**52**:211–216. DOI: 10.1095/biolreprod52.2.211

[115] Iida H, Maehara K, Doiguchi M, Mori T, Yamada F. Bisphenol A-induced apoptosis of cultured rat Sertoli cells. Reproductive Toxicology. 2003;**17**:457–464. DOI: 10.1016/s0890-6238(03)00034-0

[116] Tabuchi Y, Konodo T. cDNA microarray analysis reveals chop-10 plays a key role in Sertoli cell injury induced by bisphenol A. Biochemical and Biophysical Research Communications. 2003;**305**:54–61. DOI: 10.1016/s0006-291x(03)00708-3

[117] Toyama Y, Suzuki-Toyota F, Maekawa M, Ito C, Toshimori K. Adverse effects of bisphenol A to spermatogenesis in mice and rats. Archives of Histology and Cytology. 2004;**67**:373–381. DOI: 10.1679/aohc.67.373

[118] Fiorini C, Tilloy-Ellul A, Chevalier S, Charusel C, Pointis G. Sertoli cell junctional proteins as early targets for different classes of reproductive toxicants. Reproductive Toxicology. 2004;**18**:413–421. DOI: 10.1016/j.reprotox.2004.01.002

[119] Tainaka H, Takahashi H, Umezawa M, Tanaka H, Nishimune Y, Oshio S, Takeda K. Evaluation of the testicular toxicity of prenatal exposure to bisphenol A based on microarray analysis combined with MeSH annotation. The Journal of Toxicological Sciences. 2012;**37**:539–548. DOI: 10.2131/jts.37.539

[120] Su L, Mruk DD, Cheng CY. Drug transporters, the blood-testis barrier, and spermatogenesis. Journal of Endocrinology. 2011;**208**:207–223. DOI: 10.5353/th_b4775281

[121] Griveau JF, Dumont E, Renard P, Callegari JP, Le Lannou D. Reactive oxygen species, lipid peroxidation and enzymatic defence systems in human spermatozoa. The Journal of the Society for Reproduction and Fertility. 1995;**103**:17–26. DOI: 10.1530/jrf.0.1030017

[122] Aitken RJ, Clarkson JS, Fishel S. Generation of reactive oxygen species, lipid peroxidation, and human sperm function. Biology of Reproduction. 1989;**41**:183–197. DOI: 10.1095/biolreprod41.1.183

[123] El-Beshbishy HA, Aly HAA, El-Shafey M. Lipoic acid mitigates bisphenol A-induced testicular mitochondrial toxicity in rats. Toxicology and Industrial Health. 2013;**29**:875–887. DOI: 10.1177/0748233712446728

[124] De Flora S, Micale RT, La Maestra S, Izzotti A, D'Agostini F, Camoirano A, Davoli SA, Troglio MG, Rizzi F, Davalli P, Bettuzzi S. Upregulation of clusterin in prostate and DNA damage in spermatozoa from bisphenol A-treated rats and formation of DNA

adducts in cultured human prostatic cells. Toxicological Sciences. 2011;**122**:45–51. DOI: 10.1093/toxsci/kfr096

[125] Barbonetti A, Castellini C, Di Giammarco N, Santilli G, Francavilla S, Francavilla F. *In vitro* exposure of human spermatozoa to BPA induces pro-oxidative/apoptotic mitochondrial dysfunction. Reproductive Toxicology. 2016;**66**:61–67. DOI: 10.1016/j.reprotox.2016.09.014

[126] Atkinson A, Roy D. *In vitro* conversion of environmental estrogenic chemical bisphenol A to DNA binding metabolite(s). Biochemical and Biophysical Research Communications. 1995; **210**:424–433. DOI: 10.1006/bbrc.1995.1678

[127] Jonathan N, Steinmets R. Xenoestrogens: The emerging story of bisphenol A. Trends in Endocrinology and Metabolism. 1998;**9**:124–128. DOI: 10.1016/s1043-2760(98)00029-0

[128] Atkinson A, Roy D. *In vivo* DNA adduct formation by bisphenol A. Environmental and Molecular Mutagenesis. 1995;**26**:60–66. DOI: 10.1002/em.2850260109

[129] Knaak JB, Sullivan LJ. Metabolism of bisphenol A in the rat. Toxicology and Applied Pharmacology. 1966;**8**:175–184. DOI: 10.1016/s0041-008x(66)80001-7

[130] Tiwari D, Vanage G. Mutagenic effect of bisphenol A on adult rat male germ cells and their fertility. Reproductive Toxicology. 2013;**40**:60–68. DOI: 10.1016/j.reprotox.2013.05.013

[131] Mileva G, Baker SL, Konkle ATM, Bielajew C. Bisphenol-A: Epigenetic reprogramming and effects on reproduction and behavior. International Journal of Environmental Research and Public Health. 2014;**11**:7537–7561. DOI: 10.3390/ijerph110707537

[132] Farabollini F, Porrini S, Dessi-Fulgheri F. Perinatal exposure to the estrogenic pollutant bisphenol A affects behavior in male and female rats. Pharmacology Biochemistry and Behavior. 1999;**64**:687–694. DOI: 10.1016/s0091-3057(99)00136-7

[133] Jones BA, Watson NV. Perinatal BPA exposure demasculinizes males in measures of affect but has no effect on water maze learning in adulthood. Hormones and Behavior. 2012;**61**:605–610. DOI: 10.1016/j.yhbeh.2012.02.011

[134] Della Seta D, Minder I, Belloni V, Aloisi AM, Dessi-Fulgheri F, Farabollini F. Pubertal exposure to estrogenic chemicals affects behavior in juvenile and adult male rats. Hormones and Behavior. 2006;**50**:301–307. DOI: 10.1016/j.yhbeh.2006.03.015

[135] Manikkam M, Guerrero-Bosagna C, Tracey R, Haque M, Skinner MK. Transgenerational actions of environmental compounds on reproductive disease and identification of epigenetic biomarkers of ancestral exposures. PLoS One. 2012;**7**:31901. DOI: 10.1371/journal.pone.0031901

[136] Skinner MK, Guerrero-Bosagna C, Haque M, Nilsson E, Bhandari R, McCarrey JR. Environmentally induced transgenerational epigenetic reprogramming of primordial germ cells and the subsequent germ line. PLoS One. 2013;**8**:66318. DOI: 10.1371/journal.pone.0066318

Chapter 3

The Ovary as a Target Organ for Bisphenol A Toxicity

Anna Ptak, Marta Hoffmann and Agnieszka Rak

Additional information is available at the end of the chapter

http://dx.doi.org/10.5772/intechopen.68241

Abstract

The ovary is a hormone-sensitive organ that produces steroid hormones. Recent studies show that bisphenol A (BPA) can affect female reproduction; thus, it is important to identify the possible toxic effects of BPA on the ovary because this organ is indispensable for fertility. This chapter summarises the effects of BPA on the ovary by describing how they directly affect folliculogenesis, steroidogenesis and receptor signalling and how they indirectly affect the expression of adipokines and/or their receptors, which exert endocrine or autocrine functions within the ovary.

Keywords: bisphenol A, ovary, folliculogenesis, steroidogenesis, ovarian cancer, adipokines

1. Introduction

In the human, female germ cells develop during the first trimester of pregnancy, whereas primordial follicles develop between the second and third trimesters. Females are born with an entire lifetime supply of non-proliferating oocytes (primordial follicles) that survive for ~50 years [1]. Folliculogenesis is the process by which immature primordial follicles develop into preovulatory follicles (Graafian follicles). More than 99% of follicles never enter the preovulatory stage; instead, they undergo atresia through cell apoptosis. After ovulation, granulosa and theca cells undergo luteinisation and develop into the corpus luteum (CL). Folliculogenesis and oocyte health depend on ovarian and systemic hormones.

The cycling ovary comprises follicles and the CL. During steroidogenesis, antral follicles produce oestrogens [principally 17β-oestradiol (E_2)] from androgens [androstenedione (A4) and testosterone (T)], whereas the CL produces progesterone (P_4). This balance can be disrupted by altering the concentrations of oestrogen, androgen and/or P_4 or by affecting the expression of steroid hormone receptors. The ovarian steroid hormone receptors include those for oestrogen (ER), androgen (AR) and P_4 (PR), as well as those for luteinising hormone (LH) and follicle-stimulating hormone (FSH). The endocrine system is disrupted when a hormone can no longer bind its receptor due to a disruption in hormone synthesis or receptor binding (**Figure 1**). Additionally, a disruption in folliculogenesis or CL formation can lead to reproductive disturbances, such as aneuploidy, anovulation, decreased fertility, polycystic ovary syndrome (PCOS) and premature ovarian failure (POF). The overall damage to the ovary and its effects on fertility depends on the type of follicles affected [2].

Hormonal disturbances also underlie ovarian carcinogenesis and oestrogens, androgens, P_4, LH and FSH have been proposed to promote ovarian cancer development [3]. Depending on the cellular origin of the tumour, ovarian cancer can be classified as epithelial, stromal or germinal, with each tumour possessing different histopathological features and clinical outcomes (**Figure 2**). Epithelial cell tumours account for ~80–90% of ovarian malignancies, whereas stromal tumours account for ~8%. The most frequently diagnosed type of stromal tumour is the granulosa cell tumour (GCT).

Previous studies show correlations between women working in graphics and printing industries and increased risk of ovarian cancer [4], as well as between women working in similar industries and ovarian cancer mortality [5]. The increased incidence of ovarian cancers cannot be explained by genetic factors. We believe that environmental factors, such as toxic chemicals, can cause ovarian cancer, but it is very difficult to prove cause and effect.

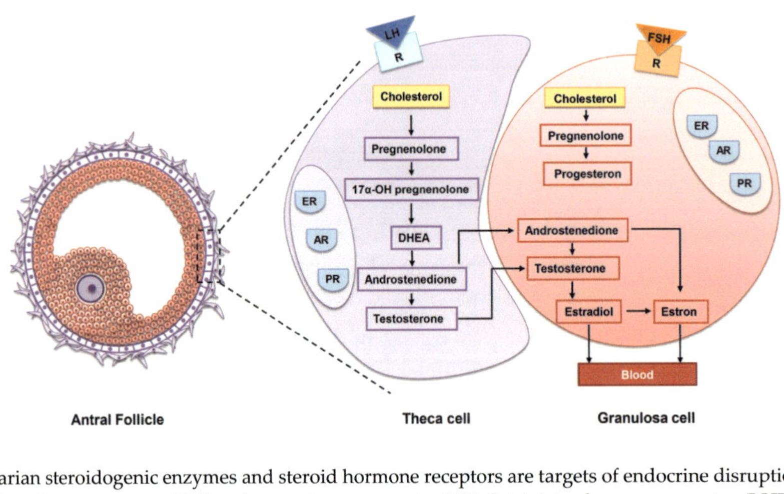

Figure 1. Ovarian steroidogenic enzymes and steroid hormone receptors are targets of endocrine disruption. Oestrogen receptor (ER), androgen receptor (AR) and progesterone receptor (PR), luteinising hormone receptor (LHR) and follicle-stimulating hormone receptor (FSHR) and dehydroepiandrosterone (DHEA).

http://dx.doi.org/10.5772/intechopen.68241

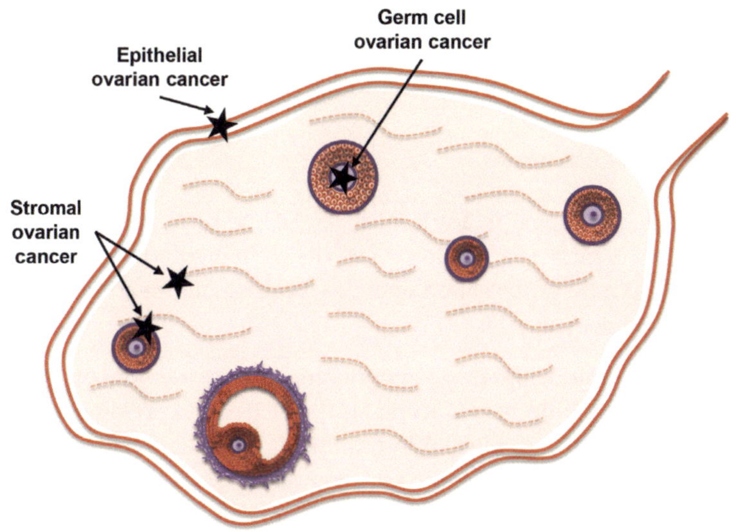

Figure 2. Different types of ovarian cancer. Stars indicate the developmental origins of the tumour.

2. Direct actions of BPA in the ovary

BPA accumulates in reproductive organs and disrupts the endocrine system. In the general population, BPA has been detected in follicular fluid at concentrations of ~1–2 ng/ml [6]. Several epidemiological studies identified correlations between BPA and various abnormalities in the ovary of foetuses and adults. Moreover, the effects of BPA in the ovary, which goes through different stages such as folliculogenesis, ovulation and luteinisation, depend on the time of exposure.

2.1. BPA action on the foetal and neonatal ovary

BPA affects oogenesis and follicle formation during foetal and early postnatal periods. For example, BPA disrupts chromosome segregation during the first meiotic division in the foetal rhesus monkey ovary. During follicle formation, BPA increases the number of multiple oocyte follicles (MOFs), which occurs when more than one oocyte is surrounded by a single layer of granulosa cells [7]. BPA also disrupts meiosis and oogenesis in the foetal mouse ovary, thereby increasing the risks of synaptic abnormalities and aneuploidy [8]. BPA also inhibits germ cell nest breakdown in the foetal mouse ovary by altering the expression of apoptotic proteins, which can lead to various fertility problems and higher percentage of dead pups [9].

Exposure of rats to BPA during the early postnatal period decreases the primordial follicle reserve and increases the incidence of MOFs [10, 11]. In the neonatal mouse ovary, BPA promotes the transition of primordial follicles to primary follicles and suppresses the meiotic maturation of oocytes due to abnormal spindle assembly during meiosis I [12]. Additionally, exposure of rats to BPA during gestational and neonatal periods induces the development of

PCOS-like syndrome during adulthood [13–15]. PCOS is the most common endocrinological pathology in women of reproductive age. It is characterised by hyperandrogenism, insulin resistance and chronic anovulation.

2.2. BPA action on the adult ovary

Oocyte abnormalities were noted in adult mice exposed to BPA, possibly due to changes in the structural integrity of microtubules that constitute meiotic spindles [16]. BPA also disrupts meiotic maturation, spindle organisation and chromosome alignment and increases oocyte degeneration in human oocytes [17].

BPA affects ovarian steroidogenesis by modulating the expression of key steroidogenic enzymes. For example, BPA decreases aromatase (*CYP19A1*) expression and E_2 production in human granulosa cells [18]. In mice, BPA inhibits P_4, testosterone (T) and E_2 synthesis by decreasing the expression of steroidogenic acute regulatory protein (*Star*), 3β-hydroxysteroid dehydrogenase (*Hsd3b1*) and 17α-hydroxylase (*Cyp17a1*) [19]. In rats, however, BPA increases P_4 and T synthesis, as well as the expression of *Star*, cholesterol side-chain cleavage enzyme (*Cyp11a1*) and *Cyp17a1*, but decreases E_2 synthesis and *Cyp19a1* expression [20]. In pigs, BPA increases basal and FSH-induced P_4 synthesis, whereas it decreases FSH-induced E_2 synthesis [21] (**Figure 3**).

In vitro studies demonstrated that BPA affects fertility by disrupting E_2 signalling, which is evolutionarily conserved among mammals and indispensable for fertility. E_2 function is mainly mediated by the classical nuclear oestrogen receptors ERα and ERβ. BPA can bind both ERα and ERβ (its affinity is higher for ERβ than ERα) [22], although its binding affinity for both receptors is greater than 1000–10000-fold lower than that for E_2 [23]. Furthermore, BPA can also induce oestrogen-like effects, because BPA elicits rapid responses through non-classical oestrogen signalling that involves the oestrogen-related receptor γ (ERRγ), [24, 25] as well as membrane-associated G protein-coupled receptor (GPR30) [26] (**Figure 3**).

Therefore, we suggest that BPA seems to be uniquely estrogenic in its receptor binding and androgenic in its hormone profile/steroidogenesis influences.

2.3. BPA and ovarian carcinogenesis

The correlation between BPA exposure and ovarian cancer is supported by little evidence. BPA exposure might increase the incidence of ovarian cysts, because women with PCOS possess higher serum BPA levels than healthy women [27]. Furthermore, women with PCOS have approximately twofold to threefold increased risk of endometrial and ovarian cancers [28, 29]. BPA might also increase the incidence of other ovarian pathologies that ultimately lead to cancer.

The balance between cell proliferation and apoptotic resistance is closely linked to cancer, and it is generally accepted as one of the major contributing factors to cancer development. BPA increases the proliferation of human epithelial ovarian cancer BG-1 [30] and OVCAR-3 [31] cells. The mitogenic effects of BPA are mainly mediated by the upregulation of genes

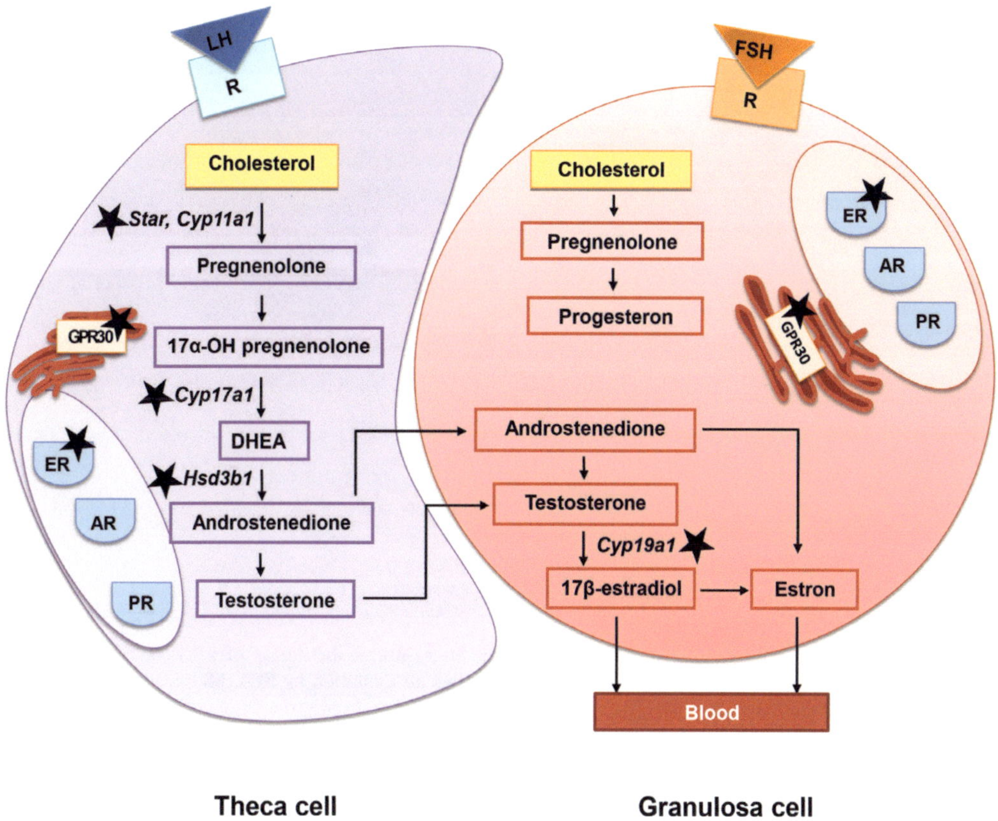

Figure 3. BPA action on ovarian steroidogenesis. Stars indicate the sites of action. Steroidogenic acute regulatory protein (Star), cholesterol side-chain cleavage enzyme (Cyp11a1), 17α-hydroxylase (Cyp17a1), 3β-hydroxysteroid dehydrogenase (Hsd3b1), aromatase (CYP19a1), oestrogen receptor (ER), androgen receptor (AR) and progesterone receptor (PR), luteinising hormone receptor (LHR) and follicle-stimulating hormone receptor (FSHR), dehydroepiandrosterone (DHEA) and membrane-associated G protein-coupled receptor (GPR30).

that induce cell proliferation (i.e., cyclin D1, cyclin A, CDK4, PCNA, E2F1 and E2F3) and the downregulation of genes that inhibit cell proliferation (i.e., p21, Weel-1 and GADD45α) in OVCAR-3 cells [31]. These findings are intriguing because decreased p21/WAF1 expression in ovarian cancer patients is an indicator of poor prognosis [32]. Furthermore, downregulation or inactivation of CDK inhibitors, such as p21Waf1/Cip1, p27Kip1 and p16Ink4a, which renders cells susceptible to extracellular signals that control proliferation, is often observed in various tumours [33]. BPA-induced cell proliferation triggers a rapid biological response involving the phosphorylation of extracellular signal-regulated kinases (ERK1/2), signal transducer and activator of transcription 3 (STAT3) and protein kinase B (AKT) in BG-1 and OVCAR-3 cells [30, 34]. BPA also inhibits OVCAR-3 cell apoptosis by activating ERK1/2 signalling [35] (**Figure 4**).

During tumourigenesis, cells can separate from the primary tumour to invade distant organs. Metastatic cancer cells undergo an epithelial-to-mesenchymal transition (EMT), which is

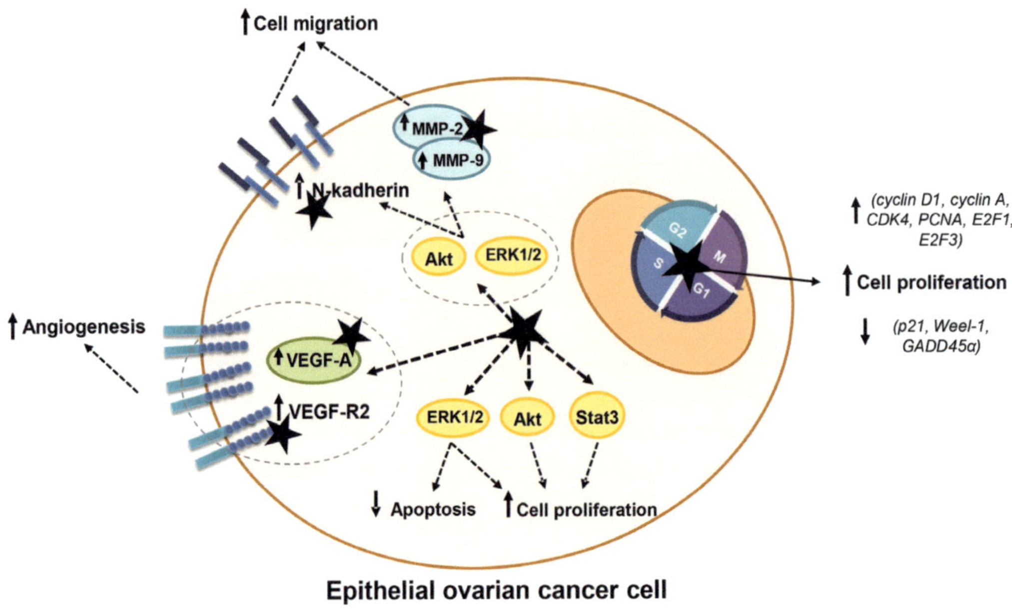

Figure 4. BPA action on epithelial ovarian cancer progression. Stars indicate the sites of BPA action. The arrow facing up indicates a stimulation, and the arrow facing down indicates an inhibition by BPA. Matrix metalloproteinase-2 (MMP-2), matrix metalloproteinase-9 (MMP-9), vascular endothelial growth factor-A (VEGF-A), vascular endothelial growth factor receptor 2 (VEGF-R2), extracellular signal-regulated kinases (ERK1/2), signal transducer and activator of transcription 3 (STAT3) and protein kinase B (Akt).

characterised by the upregulation of mesenchymal proteins such as N-cadherin, downregulation of epithelial cell-associated proteins such as E-cadherin and overexpression of matrix metalloproteinases (MMPs). MMP-2 and MMP-9 are the key enzymes required for the initial steps of ovarian cancer metastasis [36, 37]. In OVCAR-3 cells, BPA upregulates MMP-2, MMP-9 and N-cadherin expression by activating ERK1/2 and AKT signalling, which promotes cell migration [38] (**Figure 4**).

Vascular endothelial growth factor-A (VEGF-A), which is upregulated in most solid tumours, including ovarian cancers, correlates with tumour progression and poor prognosis [39, 40]. Several studies show that the serum VEGF-A level is higher in patients with ovarian cancer than in healthy individuals [41–43]. In addition, the expression of VEGF-A and its receptor (VEGF-R2) is higher in cancerous ovarian tissues than in benign or normal ovarian tissue [44]. BPA upregulates VEGF-A expression in reproductive organs, such as the uterus and vagina in the rat [45] and the ovary in the pig [46]. Moreover, BPA markedly increases VEGF-A and VEGF-R2 expression in OVCAR-3 and SKOV-3 cells [47], indicating a possible intensification of pro-angiogenic activity in ovarian cancer cells (**Figure 4**).

These findings indicate that BPA promotes the progression of epithelial ovarian cancer by stimulating epithelial cell proliferation and migration and inhibiting apoptosis. However, there is no evidence to indicate that BPA affects stromal- and germinal-derived ovarian cancers.

3. Indirect actions of BPA in the ovary through adipokines

Leptin, apelin, chemerin and adiponectin are adipokines that are mainly produced by adipose tissues, but also by other tissues. Adipokines and their receptors are expressed by cells of both the normal and cancerous ovary in humans and other mammals. They play important roles in metabolic processes, such as in the regulation of insulin sensitivity, food intake, adipogenesis and inflammation. Adipokines also regulate ovarian function, including steroidogenesis and oocyte maturation. They also affect ovarian cancer cell proliferation, apoptosis, tumour invasion and angiogenesis.

The first discovered adipokine is **leptin**, a 167-amino acid protein encoded by the *ob* gene. The leptin receptor [LEPR, also referred to as the obesity receptor (Ob-R)] is a single membrane-spanning receptor with six isoforms (Ob-Ra, b, c, d, e and f) resulting from alternative RNA splicing [48]. However, only full-length Ob-Rb can transduce signals into cells. Leptin regulates food intake, energy balance and body weight [49]. For example, there is a strong correlation between the serum leptin level and body fat content; the serum leptin level is higher in obese individuals than in those who are non-obese [50].

Granulosa and theca cells in mammalian ovaries express both leptin and LEPR. Leptin stimulates the production of ovarian steroid hormones by affecting insulin, insulin-like growth factor 1 (IGF-1) and different gonadotrophins in the cow [51–53], pig [54, 55], rodent [56, 57] and human [58–60]. There is a correlation between the serum leptin level and the P_4 concentration during the menstrual cycle in humans, as well as between E_2 and human chorionic gonadotrophin (hCG) levels throughout pregnancy [61].

Additionally, a previous study showed an association between Ob-Rb overexpression and survival in 59.2% of ovarian epithelial cell cancers [62]. OVCAR-3 cells express both long (Ob-Rb) and short (ObRt) leptin isoforms [63, 64], which associate with the progression of ovarian epithelial cell cancers. In vitro studies show that leptin promotes BG-1 and OVCAR-3 cell proliferation [34, 63] and inhibits SKOV3, MDAH2774 and OVCAR-3 cell apoptosis [34, 62]. Moreover, leptin stimulates OVCAR-3 cell migration, which is mediated via the activation of ERK1/2, AKT and STAT3 signalling [65]. Leptin also acts on ovarian cancer cells in endocrine manner because they do not produce leptin [35].

BPA can affect the expression of adipokines. BPA increases leptin mRNA expression in the preadipocyte 3T3-L1 cell line [66] and LEPR mRNA and protein expression in OVCAR-3 cells, which creates more binding sites for leptin [34] (**Figure 5**). BPA and leptin also inhibit the apoptosis of cancerous ovarian cells, indicating that BPA can potentiate leptin action in OVCAR-3 cells [35]. These results suggest that BPA increases leptin activity in cancerous ovarian cells.

Apelin is a bioactive peptide that was originally identified in bovine stomach extracts as the endogenous ligand of the orphan G protein-coupled apelin receptor (APJ) [67]. The apelin level is elevated in obese and insulin-resistant individuals and in those with high insulin levels. Apelin functions in a broad range of physiological processes, including fluid homeostasis, food intake, energy metabolism, cardiovascular function and angiogenesis.

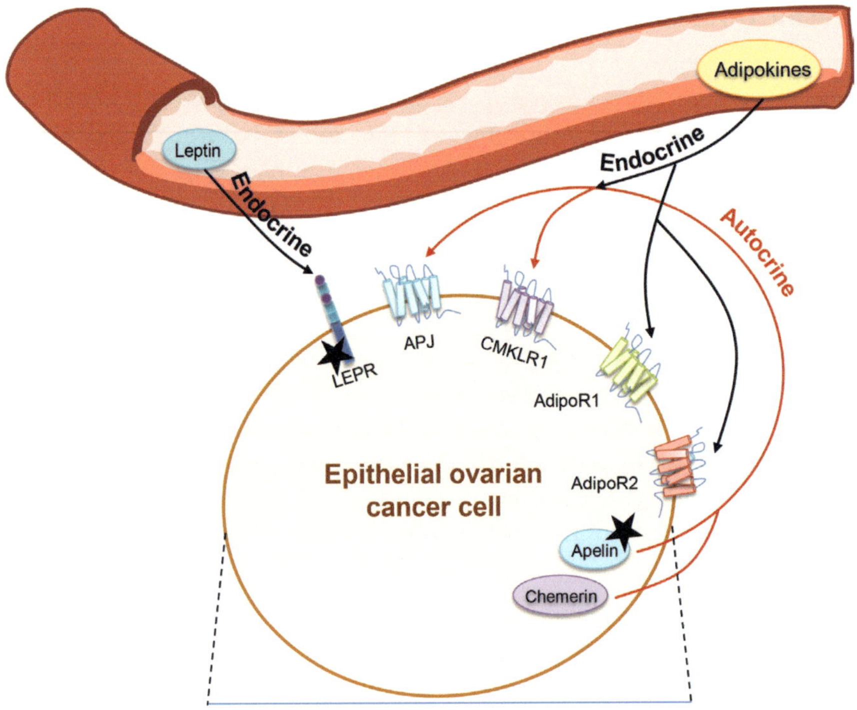

	Proliferation	Apoptosis	Migration
Leptin	↑	↓	↑
Apelin	↑	nd	nd
Chemerin	nd	nd	nd
Adiponectin	nd	nd	nd

Figure 5. BPA action on adipokines and their receptors expression in the epithelial ovarian cancer cells. Stars indicate the sites of BPA action. The arrow facing up indicates a stimulation, and the arrow facing down indicates an inhibition. Leptin receptor (LEPR), orphan G protein-coupled apelin receptor (APJ), chemokine-like receptor 1 (CMKLR1), adiponectin receptor 1 (AdipoR1) and adiponectin receptor 2 (AdipoR2).

The APJ is expressed by granulosa cells, and both apelin and its receptor are expressed by theca cells in the bovine ovary [68]. Apelin and APJ expression in theca cells are induced by LH, whereas increased APJ expression in granulosa cells associates with follicular atresia [68]. Apelin and APJ expression in mature follicles indicate that the apelin-APJ system is important for follicle selection and dominance in cows [69]. Apelin and APJ immunoexpression have been reported in granulosa and theca cells, as well as in oocytes in human follicles, at different stages of development [70]. Furthermore, apelin promotes ovarian steroid hormone secretion, in particular, P_4, and cell proliferation in pigs [71] and E_2 synthesis in humans [70], indicating that apelin has a direct role in folliculogenesis. A recent in vitro study showed that apelin stimulates rat granulosa cell proliferation; however, apelin inhibits granulosa cell apoptosis via PI3K/AKT signalling [72].

The human KGN cell line, which is derived from granulosa-like tumours, expresses apelin and APJ mRNA and protein [70]. Apelin and its receptor are also expressed by cancerous ovarian epithelial cell lines (OVCAR-3, SKOV-3 and Caov-3), the cancerous granulosa cell line (COV434) and the non-cancerous ovarian epithelial cell line (HOSEpiC). Moreover, the basal apelin concentration in both epithelial and granulosa cancer is 0.4–0.6 ng/ml. At these concentrations, apelin acts as a mitogen in these cells. However, BPA increases apelin expression and secretion only in epithelial cancer cells (**Figure 5**). BPA activates the peroxisome proliferator-activated receptor gamma (PPARγ) and not ERα and ERβ, because the PPARγ antagonist (GW9662) abolished the effects of this environmental toxicant on apelin ovarian expression [73].

Chemerin, also referred to as RARRES2 or TIG2, is secreted as prochemerin, an inactive precursor that is processed into biologically active chemerin [74]. Several isoforms of biologically active chemerin with variable C-terminal amino acids have been characterised by their abilities to bind and activate the chemokine-like receptor 1 (CMKLR1). Chemerin regulates adipogenesis, lipolysis and glucose metabolism.

Human granulosa and theca cells express chemerin and its receptor, CMKLR1. Chemerin reduces IGF-1–induced thymidine incorporation, as well as E_2 and P_4 synthesis, by decreasing the phosphorylation of the IGF-1R beta subunit and MAPK ERK1/2 in cultured human granulosa cells [75]. Similarly, chemerin decreases steroid hormone production and MAPK3/1 phosphorylation, probably through CMKLR1, in cultured bovine granulosa cells. In cumulus-oocyte complexes, chemerin blocks meiotic progression at the germinal vesicle stage and inhibits MAPK3/1 phosphorylation in both oocytes and cumulus cells during in vitro maturation [76]. Chemerin also induces rat granulosa cell apoptosis and suppresses basal, and FSH- and growth differentiation factor-9-stimulated, follicular growth in vitro [77].

Chemerin and its receptor are expressed by KGN cells, where chemerin markedly reduces IGF-1–induced cell proliferation and P_4 and E_2 synthesis [75]. Human cancerous ovarian epithelial cell lines (OVCAR-3 and SKOV-3), cancerous granulosa cell lines (COV434 and KGN) and the non-cancerous ovarian epithelial cell line (HOSEpiC) also express chemerin and its receptor. Moreover, chemerin expression decreases in BPA-treated GCTs (unpublished data). However, there is no information on the roles of chemerin in the development and progression of ovarian cancer and no data on the serum chemerin level in patients with ovarian cancer.

Adiponectin (APN), also referred to as ACRP30 or AdipoQ, is the most abundant secreted protein expressed exclusively by adipose tissue [78]. There are three major APN isoforms, namely, a trimeric low-molecular-weight (LMW) isoform, a hexameric medium-molecular -weight (MMW) isoform and a multimeric high-molecular-weight (HMW) isoform [79]. Adiponectin binds its receptors, AdipoR1 and AdipoR2.

The expression of adiponectin and its receptors has been reported in the ovary of various species, including the rat, chicken, pig, cow and human [78]. Except for the cow, adiponectin expression is absent/low in granulosa and cumulus cells of the mouse, chicken and human. In the bovine ovary, adiponectin expression varies in different cells during development [80].

Furthermore, adiponectin receptors are expressed by oocytes and early embryos of the pig and mouse [81]. In vitro studies report adiponectin to decrease insulin-induced androgen and P_4 secretion in bovine theca cells. In rat, chicken and human cultured granulosa cells, however, adiponectin increases P_4 and/or E_2 secretion in response to IGF-1. Several reports in different species, including humans, indicate that adiponectin can modulate not only granulosa cell steroidogenesis but also the expression of genes involved in ovulation. In the cow, adiponectin decreases insulin-induced steroidogenesis and increases IGF-1–induced proliferation of cultured granulosa cells. Adiponectin does not affect oocyte maturation and embryo development in vitro [82]; however, it stimulates oocyte meiotic maturation and embryo development in the pig [81].

The serum adiponectin level is markedly lower in patients with early-stage ovarian cancer than in healthy women. Adiponectin possesses anti-tumourigenic properties; it can suppress tumour growth and cell proliferation, arrest cell growth and induce apoptosis. AdipoR1 promotes KGN cell survival, whereas AdipoR2 regulates steroid hormone synthesis by activating MAPK ERK1/2 [83]. Furthermore, the AdipoR1 mRNA level was lower in Leghorn chicken cancerous ovaries than in normal ovaries [84], suggesting that adiponectin signalling restricts ovarian cancer progression by suppressing tumour cell proliferation and inducing cell apoptosis.

Human cancerous ovarian epithelial cell lines (OVCAR-3, SKOV-3 and Caov-3), the cancerous granulosa cell line (COV434) and the non-cancerous ovarian epithelial cell line (HOSEpiC) express AdipoR1 and AdipoR2, but not adiponectin. Moreover, the AdipoR1 mRNA level is markedly higher in OVCAR-3, SKOV-3, Caov-3 and COV434 cells than in HOSEpiC cells, whereas the AdipoR2 mRNA level is similar among all tested cell lines. BPA does not affect AdipoR1 and AdipoR2 expression (unpublished data), although it decreases the expression and secretion of adiponectin in 3T3-L1 adipocytes [85]. In cultured porcine ovarian follicles, however, BPA markedly increases the expression and secretion of adiponectin, as well as the expression of its receptors, indicating that this environmental toxicant contributes to ovarian dysfunction in obesity-related disorders (unpublished data).

4. Conclusion

BPA can alter ovarian function through several mechanisms. In this chapter, we have discussed two mechanisms by which BPA alters ovarian function. In the first mechanism, BPA acts directly by reducing oocyte quality after foetal and early postnatal exposure; altering the expression and/or activity of key steroidogenic enzymes required for steroid hormone synthesis; binding to steroid hormone receptors and preventing the binding of endogenous ligands; stimulating ovarian cancer cell proliferation and migration; and inhibiting cell apoptosis. In the second mechanism, BPA acts indirectly by altering the expression of adipokines and adipokine receptors, which exhibit endocrine and autocrine actions in ovarian cells. Further studies are needed to understand the effects of BPA on the ovary and its contribution to ovarian dysfunction, such as decreased fertility, PCOS and carcinogenesis.

Conflicts of interest

The authors declare no conflicts of interest.

Funding

This work was supported by the National Science Centre (NCN), Poland [grant number: DEC-2013/09/B/NZ7/00405].

Author details

Anna Ptak*, Marta Hoffmann and Agnieszka Rak

*Address all correspondence to: anna.ptak@uj.edu.pl

Department of Physiology and Toxicology of Reproduction, Institute of Zoology, Jagiellonian University in Krakow, Krakow, Poland

References

[1] Oktem O, Urman B. Understanding follicle growth in vivo. Hum Reprod. 2010;25:2944–54.

[2] Drummond AE, Britt KL, Dyson M, Jones ME, Kerr JB, O'Donnell L, Simpson ER, Findlay JK. Ovarian steroid receptors and their role in ovarian function. Mol Cell Endocrinol. 2002;191:27–33.

[3] Lukanova A, Kaaks R. Endogenous hormones and ovarian cancer: epidemiology and current hypotheses. Cancer Epidemiol Biomarkers Prev. 2005;14:98–107.

[4] Shen N, Weiderpass E, Antilla A, Goldberg MS, Vasama-Neuvonen KM, Boffetta P, Vainio HU, Partanen TJ. Epidemiology of occupational and environmental risk factors related to ovarian cancer. Scand J Work Environ Health. 1998;24:175–82.

[5] García-Pérez J, Lope V, López-Abente G, González-Sánchez M, Fernández-Navarro P. Ovarian cancer mortality and industrial pollution. Environ Pollut. 2015;205:103–10.

[6] Ikezuki Y, Tsutsumi O, Takai Y, Kamei Y, Taketani Y. Determination of bisphenol A concentrations in human biological fluids reveals significant early prenatal exposure. Hum Reprod. 2002;17:2839–41.

[7] Hunt PA, Lawson C, Gieske M, Murdoch B, Smith H, Marre A, Hassold T, VandeVoort CA. Bisphenol A alters early oogenesis and follicle formation in the fetal ovary of the rhesus monkey. Proc Natl Acad Sci U S A. 2012;109:17525–30.

[8] Susiarjo M, Hassold TJ, Freeman E, Hunt PA. Bisphenol A exposure in utero disrupts early oogenesis in the mouse. PLoS Genet. 2007;3(1):e5.

[9] Wang W, Hafner KS, Flaws JA. In utero bisphenol A exposure disrupts germ cell nest breakdown and reduces fertility with age in the mouse. Toxicol Appl Pharmacol. 2014;276:157–64.

[10] Rodríguez HA, Santambrosio N, Santamaría CG, Muñoz-de-Toro M, Luque EH. Neonatal exposure to bisphenol A reduces the pool of primordial follicles in the rat ovary. Reprod Toxicol. 2010;30:550–7.

[11] Karavan JR, Pepling ME. Effects of estrogenic compounds on neonatal oocyte development. Reprod Toxicol. 2012;34:51–6.

[12] Chao HH, Zhang XF, Chen B, Pan B, Zhang LJ, Li L, Sun XF, Shi QH, Shen W. Bisphenol A exposure modifies methylation of imprinted genes in mouse oocytes via the estrogen receptor signaling pathway. Histochem Cell Biol. 2012;137:249–59.

[13] Kato H, Ota T, Furuhashi T, Ohta Y, Iguchi T. Changes in reproductive organs of female rats treated with bisphenol A during the neonatal period. Reprod Toxicol. 200317:283–8.

[14] Fernández M, Bourguignon N, Lux-Lantos V, Libertun C. Neonatal exposure to bisphenol a and reproductive and endocrine alterations resembling the polycystic ovarian syndrome in adult rats. Environ Health Perspect. 2010;118:1217–22.

[15] Aghajanova L, Giudice LC. Effect of bisphenol A on human endometrial stromal fibroblasts in vitro. Reprod Biomed Online. 2011;22:249–56.

[16] Can A, Semiz O, Cinar O. Bisphenol-A induces cell cycle delay and alters centrosome and spindle microtubular organization in oocytes during meiosis. Mol Hum Reprod. 2005;11:389–96.

[17] Machtinger R, Combelles CM, Missmer SA, Correia KF, Williams P, Hauser R, Racowsky C. Bisphenol-A and human oocyte maturation in vitro. Hum Reprod. 2013;28:2735–45.

[18] Kwintkiewicz J, Nishi Y, Yanase T, Giudice LC. Peroxisome proliferator-activated receptor-gamma mediates bisphenol A inhibition of FSH-stimulated IGF-1, aromatase, and estradiol in human granulosa cells. Environ Health Perspect. 2010;118:400–6.

[19] Peretz J, Gupta RK, Singh J, Hernández-Ochoa I, Flaws JA. Bisphenol A impairs follicle growth, inhibits steroidogenesis, and downregulates rate-limiting enzymes in the estradiol biosynthesis pathway. Toxicol Sci. 2011;119:209–17.

[20] Zhou W, Liu J, Liao L, Han S, Liu J. Effect of bisphenol A on steroid hormone production in rat ovarian theca-interstitial and granulosa cells. Mol Cell Endocrinol. 2008;283:12–8

[21] Mlynarcíková A, Kolena J, Ficková M, Scsuková S. Alterations in steroid hormone production by porcine ovarian granulosa cells caused by bisphenol A and bisphenol A dimethacrylate. Mol Cell Endocrinol. 2005;244:57–62.

[22] Matthews JB, Twomey K, Zacharewski TR. In vitro and in vivo interactions of bisphenol A and its metabolite, bisphenol A glucuronide, with estrogen receptors alpha and beta. Chem Res Toxicol. 2001;14:149–57.

[23] Fang Fang H, Tong W, Perkins R, Soto AM, Prechtl NV, Sheehan DM. Quantitative comparisons of in vitro assays for estrogenic activities. Environ Health Perspect. 2000;108:723–9.

[24] Takayanagi S, Tokunaga T, Liu X, Okada H, Matsushima A, Shimohigashi Y. Endocrine disruptor bisphenol A strongly binds to human estrogen-related receptor gamma (ERRgamma) with high constitutive activity. Toxicol Lett. 2006;167:95–105.

[25] Matsushima A, Teramoto T, Okada H, Liu X, Tokunaga T, Kakuta Y, Shimohigashi Y. ERRgamma tethers strongly bisphenol A and 4-alpha-cumylphenol in an induced-fit manner. Biochem Biophys Res Commun. 2008;373:408–13.

[26] Thomas P, Dong J. Binding and activation of the seven-transmembrane estrogen receptor GPR30 by environmental estrogens: a potential novel mechanism of endocrine disruption. J Steroid Biochem Mol Biol. 2006;102:175–9.

[27] Kandaraki E, Chatzigeorgiou A, Livadas S, Palioura E, Economou F, Koutsilieris M, Palimeri S, Panidis D, Diamanti-Kandarakis E. Endocrine disruptors and polycystic ovary syndrome (PCOS): elevated serum levels of bisphenol A in women with PCOS. J Clin Endocrinol Metab. 2011;96(3):E480–4.

[28] Schildkraut JM, Schwingl PJ, Bastos E, Evanoff A, Hughes C. Epithelial ovarian cancer risk among women with polycystic ovary syndrome. Obstet Gynecol. 1996;88:554–9.

[29] Dumesic DA, Richards JS. Ontogeny of the ovary in polycystic ovary syndrome. Fertil Steril. 2013;100:23–38.

[30] Park SH, Kim KY, An BS, Choi JH, Jeung EB, Leung PC, Choi KC. Cell growth of ovarian cancer cells is stimulated by xenoestrogens through an estrogen-dependent pathway, but their stimulation of cell growth appears not to be involved in the activation of the mitogen-activated protein kinases ERK-1 and p38. J Reprod Dev. 2009;55:23–9.

[31] Ptak A, Wróbel A, Gregoraszczuk EL. Effect of bisphenol-A on the expression of selected genes involved in cell cycle and apoptosis in the OVCAR-3 cell line. Toxicol Lett. 2011;202:30–5.

[32] Bali A, O'Brien PM, Edwards LS, Sutherland RL, Hacker NF, Henshall SM. Cyclin D1, p53, and p21Waf1/Cip1 expression is predictive of poor clinical outcome in serous epithelial ovarian cancer. Clin Cancer Res. 2004;10:5168–77.

[33] Sherr CJ, Roberts JM. CDK inhibitors: positive and negative regulators of G1-phase progression. Genes Dev. 1999;13:1501–12.

[34] Ptak A, Gregoraszczuk EL. Bisphenol A induces leptin receptor expression, creating more binding sites for leptin, and activates the JAK/Stat, MAPK/ERK and PI3K/Akt signalling pathways in human ovarian cancer cell. Toxicol Lett. 2012;210:332–7

[35] Ptak A, Rak-Mardyła A, Gregoraszczuk EL. Cooperation of bisphenol A and leptin in inhibition of caspase-3 expression and activity in OVCAR-3 ovarian cancer cells. Toxicol In Vitro. 2013;27:1937–43.

[36] Symowicz J, Adley BP, Gleason KJ, Johnson JJ, Ghosh S, Fishman DA, Hudson LG, Stack MS. Engagement of collagen-binding integrins promotes matrix metalloproteinase-9-dependent E-cadherin ectodomain shedding in ovarian carcinoma cells. Cancer Res. 2007;67:2030–9.

[37] Kenny HA, Lengyel E. MMP-2 functions as an early response protein in ovarian cancer metastasis. Cell Cycle. 2009;8:683–8.

[38] Ptak A, Hoffmann M, Gruca I, Barć J. Bisphenol A induce ovarian cancer cell migration via the MAPK and PI3K/Akt signalling pathways. Toxicol Lett. 2014;229:357–65.

[39] Siddiqui GK, Elmasry K, Wong Te Fong AC, Perrett C, Morris R, Crow JC, Maclean AB. Prognostic significance of intratumoral vascular endothelial growth factor as a marker of tumour angiogenesis in epithelial ovarian cancer. Eur J Gynaecol Oncol. 2010;31:156–9.

[40] Yu L, Deng L, Li J, Zhang Y, Hu L. The prognostic value of vascular endothelial growth factor in ovarian cancer: a systematic review and meta-analysis. Gynecol Oncol. 2013;128:391–6.

[41] Paley PJ, Staskus KA, Gebhard K, Mohanraj D, Twiggs LB, Carson LF, Ramakrishnan S. Vascular endothelial growth factor expression in early stageovarian carcinoma. Cancer. 1997;80:98–106.

[42] Yamamoto S, Konishi I, Mandai M, Kuroda H, Komatsu T, Nanbu K, Sakahara H, Mori T. Expression of vascular endothelial growth factor (VEGF) in epithelial ovarian neoplasms: correlation with clinicopathology and patient survival, and analysis of serum VEGF levels. Br J Cancer. 1997;76:1221–7.

[43] Kraft A, Weindel K, Ochs A, Marth C, Zmija J, Schumacher P, Unger C, Marmé D, Gastl G. Vascular endothelial growth factor in the sera and effusions of patients with malignant and nonmalignant disease. Cancer. 1999;85:178–87.

[44] Chen H, Ye D, Xie X, Chen B, Lu W. VEGF, VEGFRs expressions and activated STATs in ovarian epithelial carcinoma. Gynecol Oncol. 2004;94:630–5.

[45] Long X, Burke KA, Bigsby RM, Nephew KP. Effects of the xenoestrogen bisphenol A on expression of vascular endothelial growth factor (VEGF) in the rat. Exp Biol Med (Maywood). 2001;226:477–83.

[46] Grasselli F, Baratta L, Baioni L, Bussolati S, Ramoni R, Grolli S, Basini G. Bisphenol A disrupts granulosa cell function. Domest Anim Endocrinol. 2010;39:34–9.

[47] Ptak A, Gregoraszczuk EL. Effects of bisphenol A and 17β-estradiol on vascular endothelial growth factor A and its receptor expression in the non-cancer and cancer ovarian cell lines. Cell Biol Toxicol. 2015;31:187–97.

[48] Gorska E, Popko K, Stelmaszczyk-Emmel A, Ciepiela O, Kucharska A, Wasik M. Leptin receptors. Eur J Med Res. 2010;15:50–4.

[49] Morris DL, Rui L. Recent advances in understanding leptin signaling and leptin resistance. Am J Physiol Endocrinol Metab. 2009;297:E1247–59.

[50] Grossmann ME, Ray A, Nkhata KJ, Malakhov DA, Rogozina OP, Dogan S, Cleary MP. Obesity and breast cancer: status of leptin and adiponectin in pathological processes. Cancer Metastasis Rev. 2010;29:641–53.

[51] Spicer LJ, Francisco CC. The adipose obese gene product, leptin: evidence of a direct inhibitory role in ovarian function. Endocrinology. 1997;138:3374–9.

[52] Spicer LJ, Francisco CC. Adipose obese gene product, leptin, inhibits bovine ovarian thecal cell steroidogenesis. Biol Reprod. 1998;58:207–12.

[53] Spicer LJ, Chamberlain CS, Francisco CC. Ovarian action of leptin: effects on insulin-like growth factor-I-stimulated function of granulosa and thecal cells. Endocrine. 2000;12:53–9.

[54] Gregoraszczuk EŁ, Wójtowicz AK, Ptak A, Nowak K. In vitro effect of leptin on steroids' secretion by FSH- and LH-treated porcine small, medium and large preovulatory follicles. Reprod Biol. 2003;3:227–39.

[55] Gregoraszczuk EL, Ptak A, Wojtowicz AK, Gorska T, Nowak KW. Estrus cycle-dependent action of leptin on basal and GH or IGF-I stimulated steroid secretion by whole porcine follicles. Endocr Regul. 2004;38:15–21.

[56] Zachow RJ, Magoffin DA. Direct intraovarian effects of leptin: impairment of the synergistic action of insulin-like growth factor-I on follicle-stimulating hormone-dependent estradiol-17 beta production by rat ovarian granulosa cells. Endocrinology. 1997;138:847–50.

[57] Duggal PS, Ryan NK, Van der Hoek KH, Ritter LJ, Armstrong DT, Magoffin DA, Norman RJ. Effects of leptin administration and feed restriction on thecal leucocytes in the preovulatory rat ovary and the effects of leptin on meiotic maturation, granulosa cell proliferation, steroid hormone and PGE2 release in cultured rat ovarian follicles. Reproduction. 2002;123:891–8.

[58] Agarwal SK, Vogel K, Weitsman SR, Magoffin DA. Leptin antagonizes the insulin-like growth factor-I augmentation of steroidogenesis in granulosa and theca cells of the human ovary. J Clin Endocrinol Metab. 1999;84:1072–6.

[59] Guo X, Chen S, Xing F. Effects of leptin on estradiol and progesterone production by human luteinized granulosa cells in vitro. Zhonghua Fu Chan Ke Za Zhi. 2001;36:95–7.

[60] Tsai EM, Yang CH, Chen SC, Liu YH, Chen HS, Hsu SC, Lee JN. Leptin affects pregnancy outcome of in vitro fertilization and steroidogenesis of human granulosa cells. J Assist Reprod Genet. 2002;19:169–76.

[61] Hardie L, Trayhurn P, Abramovich D, Fowler P. Circulating leptin in women: a longitudinal study in the menstrual cycle and during pregnancy. Clin Endocrinol (Oxf). 1997;47:101–6.

[62] Uddin S, Bu R, Ahmed M, Abubaker J, Al-Dayel F, Bavi P, Al-Kuraya KS. Overexpression of leptin receptor predicts an unfavorable outcome in Middle Eastern ovarian cancer. Mol Cancer. 2009;8:74.

[63] Choi JH, Park SH, Leung PC, Choi KC. Expression of leptin receptors and potential effects of leptin on the cell growth and activation of mitogen-activated protein kinases in ovarian cancer cells. J Clin Endocrinol Metab. 2005;90:207–10.

[64] Ptak A, Kolaczkowska E, Gregoraszczuk EL. Leptin stimulation of cell cycle and inhibition of apoptosis gene and protein expression in OVCAR-3 ovarian cancer cells. Endocrine. 2013;43:394–403.

[65] Hoffmann M, Fiedor E, Ptak A. 17β-Estradiol Reverses Leptin-Inducing Ovarian Cancer Cell Migration by the PI3K/Akt Signaling Pathway. Reprod Sci. 2016;23:1600–1608.

[66] Phrakonkham P, Viengchareun S, Belloir C, Lombès M, Artur Y, Canivenc-Lavier MC. Dietary xenoestrogens differentially impair 3T3-L1 preadipocyte differentiation and persistently affect leptin synthesis. J Steroid Biochem Mol Biol. 2008;110:95–103.

[67] Tatemoto K, Hosoya M, Habata Y, Fujii R, Kakegawa T, Zou MX, Kawamata Y, Fukusumi S, Hinuma S, Kitada C, Kurokawa T, Onda H, Fujino M. Isolation and characterization of a novel endogenous peptide ligand for the human APJ receptor. Biochem Biophys Res Commun. 1998;251:471–6.

[68] Shimizu T, Kosaka N, Murayama C, Tetsuka M, Miyamoto A. Apelin and APJ receptor expression in granulosa and theca cells during different stages of follicular development in the bovine ovary: Involvement of apoptosis and hormonal regulation. Anim Reprod Sci. 2009;116:28–37.

[69] Schilffarth S, Antoni B, Schams D, Meyer HH, Berisha B. The expression of apelin and its receptor APJ during different physiological stages in the bovine ovary. Int J Biol Sci. 2009;5:344–50.

[70] Roche J, Ramé C, Reverchon M, Mellouk N, Cornuau M, Guerif F, Froment P, Dupont J. Apelin (APLN) and apelin receptor (APLNR) in human ovary: Expression, signaling, and regulation of steroidogenesis in primary human Luteinized Granulosa Cells. Biol Reprod. 2016;95:104.

[71] Rak A, Drwal E, Rame C, Knapczyk-Stwora K, Słomczyńska M, Dupont J, Gregoraszczuk E. Apelin and apelin receptor (APJ) are expressed in porcine ovarian follicles and regulate steroidogenesis and proliferation through APJ activation and different signaling pathways in co-culture of granulosa and theca cells. Reproductive Sciences, 2017, in press.

[72] Shuang L, Jidong W, Hongjuan P, Zhenwei Y. Effects of apelin on proliferation and apoptosis in rat ovarian granulosa cells. Clin Exp Obstet Gynecol. 2016;43:409–13.

[73] Hoffmann M, Fiedor E, Ptak A. Bisphenol A and its derivatives tetrabromobisphenol A and tetrachlorobisphenol A induce apelin expression and secretion in ovarian cancer cells through a peroxisome proliferator-activated receptor gamma-dependent mechanism. Toxicol Lett. 2017, doi:10.1016/j.toxlet.2017.01.006

[74] Du XY, Leung LL. Proteolytic regulatory mechanism of chemerin bioactivity. Acta Biochim Biophys Sin (Shanghai). 2009;41:973–9.

[75] Reverchon M, Cornuau M, Ramé C, Guerif F, Royère D, Dupont J. Chemerin inhibits IGF-1-induced progesterone and estradiol secretion in human granulosa cells. Hum Reprod. 2012;27:1790–800.

[76] Reverchon M, Bertoldo MJ, Ramé C, Froment P, Dupont J. CHEMERIN (RARRES2) decreases in vitro granulosa cell steroidogenesis and blocks oocyte meiotic progression in bovine species. Biol Reprod. 2014;90:102.

[77] Kim JY, Xue K, Cao M, Wang Q, Liu JY, Leader A, Han JY, Tsang BK. Chemerin suppresses ovarian follicular development and its potential involvement in follicular arrest in rats treated chronically with dihydrotestosterone. Endocrinology. 2013;154:2912–23.

[78] Campos DB, Palin MF, Bordignon V, Murphy BD. The 'beneficial' adipokines in reproduction and fertility. Int J Obes (Lond). 2008;32:223–31.

[79] Kadowaki T, Yamauchi T. Adiponectin and adiponectin receptors. Endocr Rev. 2005;26: 439–51.

[80] Tabandeh MR, Hosseini A, Saeb M, Kafi M, Saeb S. Changes in the gene expression of adiponectin and adiponectin receptors (AdipoR1 and AdipoR2) in ovarian follicular cells of dairy cow at different stages of development. Theriogenology. 2010;73:659–69.

[81] Chappaz E, Albornoz MS, Campos D, Che L, Palin MF, Murphy BD, Bordignon V. Adiponectin enhances in vitro development of swine embryos. Domest Anim Endocrinol. 2008;35:198–207.

[82] Maillard V, Uzbekova S, Guignot F, Perreau C, Ramé C, Coyral-Castel S, Dupont J. Effect of adiponectin on bovine granulosa cell steroidogenesis, oocyte maturation and embryo development. Reprod Biol Endocrinol. 2010;8:23.

[83] Pierre P, Froment P, Nègre D, Ramé C, Barateau V, Chabrolle C, Lecomte P, Dupont J. Role of adiponectin receptors, AdipoR1 and AdipoR2, in the steroidogenesis of the human granulosa tumor cell line, KGN. Hum Reprod. 2009;24:2890–901.

[84] Ocon-Grove O, Krzysik-Walker S, Giles J, Johnson P, Hendricks G, Ramachandran R. Expression of adiponectin and its receptors, AdipoR1, AdipoR2 and T-cadherin, in chicken ovarian tumors. Biol Reprod. 2008;78:139–140.

[85] Kidani T, Kamei S, Miyawaki J, Aizawa J, Sakayama K, Masuno H. Bisphenol A downregulates Akt signaling and inhibits adiponectin production and secretion in 3T3-L1 adipocytes. J Atheroscler Thromb. 2010;17:834–43.

Chapter 4

Bisphenol A in Chronic Kidney Disease

Giuseppe Palladino and Luisa Sereni

Additional information is available at the end of the chapter

http://dx.doi.org/10.5772/intechopen.68681

Abstract

Several types of medical devices are produced using polycarbonate (PC) polymer. Unfortunately, these medical devices produced using PC could contain and release Bisphenol A (BPA) residual in routinely use. Published evidence on BPA in dialysis or Chronic Kidney patients (CKD) is scarce and limited to the observation of increased blood BPA levels. Increased serum BPA with decreasing renal function was observed in a smaller study of 32 CKD patients, suggesting that BPA may accumulate. Recently a crossover study evaluated the impact of the choice of dialyzer (BPA-free versus BPA-containing) on serum and intracellular BPA levels and on inflammation and oxidative stress markers. Currently, BPA is still considered, from regulatory agencies, safe enough in the general population, despite several red flags, as it is readily excreted in the urine. However, patients in End Stage Renal Disease (ESRD) are unable to excrete BPA in their urine, leading to BPA accumulation. Repeated loading of BPA during hemodialysis with BPA-containing membranes may aggravate the problem due to migration of BPA from dialyzers to the blood of patients. In contrast, some recent studies on the chronic use of BPA-free dialyzers, results in decreased BPA levels.

Keywords: BPA, chronic kidney disease, hemodialysis

1. Introduction

Bisphenol A, normally abbreviated as BPA, is an organic compound with two vicinal phenolic group. It is also known as 2,2-bis-(4-hydroxyphenyl) propane. BPA is a high-volume industrial chemical used in the production of epoxy resins and polycarbonate (PC) plastics.

The chemical synthesis and industrial conversion of BPA in PC and epoxy resin are shown in **Figure 1**.

Many food and drink containers are manufactured with PC plastics. On the other side, epoxy resins are normally used as inner liners in food and drink containers; polycarbonate plastics may be encountered in many products, especially in food and drink containers, while epoxy resins are frequently used as inner liners of metallic food and drink recipients with the aim to prevent corrosion. Some thermal paper used in cash registers or similar devices could be a source of BPA. Additionally, BPAs have been used in polyvinylchloride (PVC) industries and metal foundries for cast and molding production.

BPA was synthesized for the first time by the Russian chemist A.P. Dianin in 1891, and in the early 1930s, the British biochemist E.C. Dodds tested BPA as an artificial estrogen but found it less effective than estradiol [1].

Dodds eventually developed a structurally similar compound, diethylstilbestrol (DES), which was used as a synthetic estrogen drug [2] in women and in animals until it was banned in 1971 due to its risk of causing cancer. Actually, bisphenol A is used primarily to make plastics, and BPA is contained in products that have been in commercial use since 1957. BPA-based plastic is clear and tough and is made into a variety of common consumer goods, such as water bottles, sports equipment, CDs, and DVDs. Epoxy resins containing BPA are used to line water pipes, as coatings on the inside of many food and beverage cans and in making thermal paper such as that used in sale receipts. In 2015, an estimated 5.4 million tonnes of BPA chemical were produced for manufacturing polycarbonate plastic, making it one of the highest volume of chemicals produced worldwide.

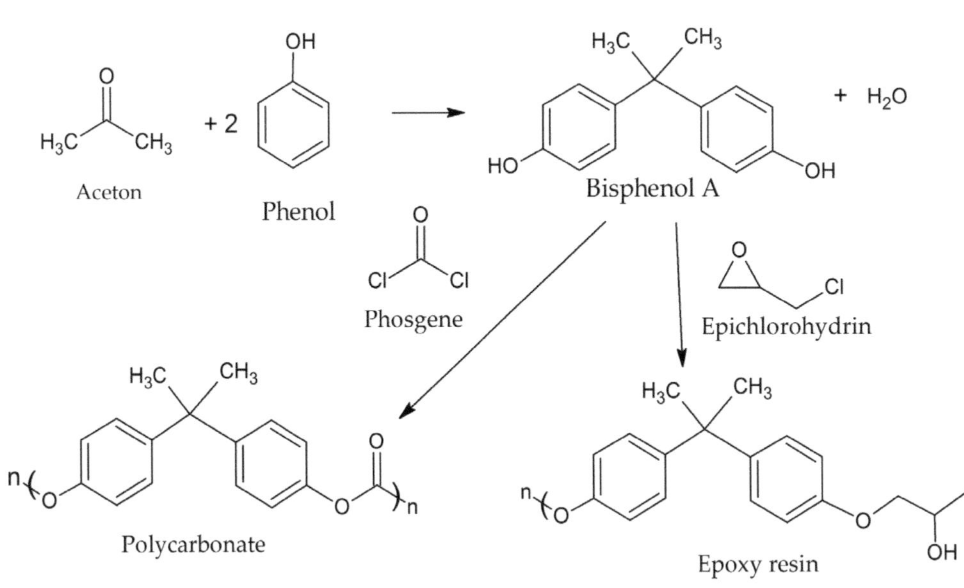

Figure 1. Bisphenol A as a commodity chemical and essential component of two classes of polymers.

2. Human exposure to BPA

The human population is primarily exposed to BPA through the diet. Other possible sources of bisphenol A exposure are air, dust, water, and skin contact with thermal paper. From this point of view, most of 50% of BPA exposure account from food and beverages packed and distributed in boxes, bottles, or cans containing BPA. The remaining 50% is coming from thermal paper, or paper in general, in contact with skin. Monomers of BPA are released from slow decay of polymers in contact with food and liquids. BPA release could be accelerated by heating, contact with alkaline or acidic substances, repeated use and exposure to microwaves. In **Table 1**, an overview of typical concentration in food and non-food BPA-containing materials is shown.

BPA may be absorbed in the gastrointestinal tract after ingesting products packed in plastic containers. Like intestinal phenols, BPA is conjugated by glucuronic acid in bowel and liver and excreted in urine as BPA-glucuronide [3].

Levels, normally less than 1 μg/L, measured in human biological fluids indicate a recent exposure to the molecule because of its rapid conjugation and elimination by the liver and gastrointestinal tract in a few hours. Kinetics in vivo study support this hypothesis of rapid plasma clearance of BPA metabolites.

Measured concentrations of BPA in human blood, urine, and other tissues indicated that the majority of the population (91–99%) has detectable levels of BPA-conjugates in their urine, confirming that exposure is widespread in the human population.

Type of food product	Typical BPA concentration
Canned food	30–50 μg/kg
Noncanned food	0–10 μg/kg
Canned drinks and dairy products	0.5–5 μg/kg
Noncanned drinks and dairy products	0–1 μg/kg
Initial breast milk	3 μg/L
Mature breast milk	1.5 μg/L
Cosmetics	31 μg/kg product
Thermal paper	0.8–3.2 μg/100 g
Indoor air	0.5–5.3 ng/m^3
Dust	117–20,000 μg/kg
Toys/rattles (mouthed)	0.14 μg/kg product
Pacifiers (mouthed)	0.28–0.36 μg/product

Source: European Food Safety Authority.

Table 1. Overview of BPA typical concentration in food and nonfood BPA-containing materials.

The largest scale studies with a consistently high number of enrolled participants (n = 2517 and 5476 individuals) spread over a broad range of age were carried out in the USA and Canada, respectively [4–6].

In these studies, the highest BPA levels detected in urine were 3.6 ng/mL for the US and 1.30 ng/mL for the Canadian ones in a subgroup of population within the age group of 6 and 11 years. On the contrary, the adult population had lower BPA urine concentration: 2.6 and 1.16 ng/mL, respectively. Zhang et al. show the same results in a recent study conducted on the Asian population [7].

Mose et al. [8] studied the BPA trans-placental transfer rate in human placentas in ex vivo experiments. Results led the authors to conclude that free BPA can cross the placenta by passive diffusion with a trans-placental transfer rate of 1 (e.g., the concentration in the fetal blood was equal to the concentration in the blood of the mother), as previously demonstrated by Balakrishnan et al. [9].

3. Bisphenol A and human health

Searching in the literature an increasing number of studies are found containing data coming from epidemiological studies on the association between BPA exposure and healthy outcomes. Unfortunately, this number is still limited, and results are coming from cross-sectional trials that limit their interpretability for pathology with long latency periods like cardiovascular diseases or diabetes. An example of number of publication per year, found in PubMed, is shown in **Figure 2**.

Six cross-sectional analyses of data from the US National Health and Nutrition Examination Survey (NHANES) reported associations of BPA exposure with self-reported diagnosis of pre-existing cardiovascular disease, hypertension, obesity, diabetes, and liver-enzyme abnormalities [10–15]. Two other studies in the US [16] and China [17] reported an association between BPA exposure and coronary disease at the time of diagnosis and obesity and insulin resistance, respectively. In addition, a study found associations between urine BPA and immune function and allergy [18]. These cross-sectional analyses have the same weaknesses that limit their interpretation. One of the major limitations of these studies could be assigned to the problem relating to sampling procedure (single spot urine) reflecting only recent BPA exposure and not on a long period (months or years) much more useful to assess the exposure effect on cardiovascular disease and diabetes pathologies.

Progressive exposure to BPA can affect adiposity, glucose or insulin regulation, lipid profiles or other end-points relating to diabetes or metabolic syndrome [19–27].

Finally, BPA could have a negative effect on the heart: stimulating estrogen concentration and modifying free calcium concentration control inside heart cells in women. Provoking an increase of Ca^{2+} release from sarcoplasmic reticule that could cause arrhythmias that in some case could degenerate into infarction [28].

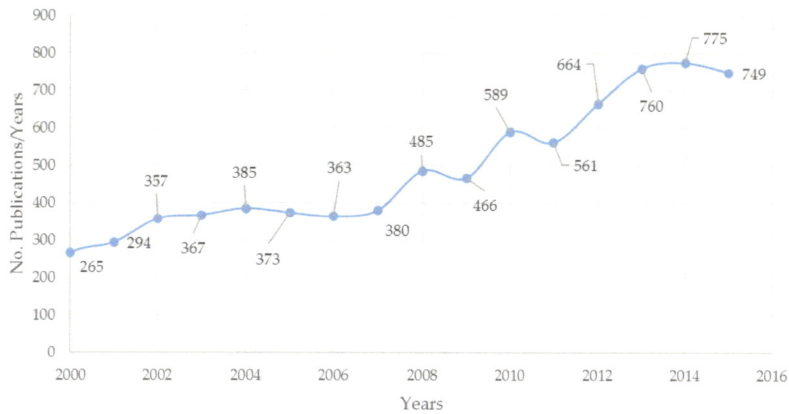

Figure 2. Number of publication on BPA (2000–2015). Source: PubMed.

4. Bisphenol A in hemodialyzers

Several types of medical devices are produced using polycarbonate (PC) polymer. Industries utilize PC for its toughness and stability, optical clarity, and resistance to heat and electricity. Unfortunately, these medical devices produced using PC could contain and release BPA residual in routine use. Additional source of BPA is coming from other materials such as dental supplies manufactured using bisphenol derivatives like bisphenol A glycidyl methacrylate (Bis-GMA) and bisphenol A dimethacrylate (Bis-DMA). Bisphenol A is also used in the production of inks and adhesives, as well as in polysulphone (PS) membranes widely used in hemodialyzer production.

Recently, the Scientific Committee on Emerging and Newly Identified Health Risks (SCENIHR) published an opinion of the safety of bisphenol A use in medical devices [29].

In their report, the SCENIHR described the risk assessment of exposure to BPA via medical devices that are manufactured with materials that potentially leach BPA. This include oral (via dental material), subcutaneous, and intravenous (e.g., during hemodialysis) routes of exposure.

This group evaluated the different scenarios of exposition considering the type of materials used, frequency and duration of a single treatment as well as the information relating to BPA leaching, generating an accurate report on short- and long-term toxicological exposure.

As a conclusion, the SCENIHR group reports a possible existence of adverse events risk of BPA when this is directly available in the blood, and in a particular case, for neonates in intensive care units, infants undergoing prolonged medical procedures, and for dialysis patients. Although the use of medical device should be considered together with the benefit for the patient, a BPA-free device should be taken into consideration if available. What they also suggest is to consider the possibility of replacing BPA in medical device products against their efficiency in the treatment, as well as the toxicological profile of the alternative materials.

Several studies have reported the leaching of BPA from hemodialyzers. Haishima et al. [30] studied the amount of BPA released from different hemodialyzers composed of a combination of polycarbonate housing and cellulose acetate hollow-fibers, polycarbonate housing and polysulfone fibers, and polystyrene and polysulfone. Water and bovine serum were circulated at room temperature in the four different devices tested. The bovine serum was used as a stimulant for human blood circulating into hollow fibers during hemodialysis.

BPA recovered ranged from 3.78 to 141.8 ng/module using water circulation and from 140.7 to 2090 ng/module when bovine serum was used. The highest values of BPA released corresponded to hemodialyzers consisting of PC housing and PS.

Murakami et al. [31] reported a BPA concentration of 8.33 and 12.25 ng extracted from 10 mg of polysulfone and polyester polymer alloy (PEPA) hollow fibers. The fibers, taken from individual dialyzers, were crushed and dissolved in hexane.

Fink [32] investigated BPA leaching from five different types of dialyzers and polyvinylchloride blood tubing. All the dialyzers were composed of either polycarbonate or polysulfone: polycarbonate housing and polysulfone-polyvinylpyrrolidone (PVP) blend membranes, PC housing and polyamide-polysulfone blend membrane, and polypropylene housing, polysulfone-PVP blend membrane. Different surface area range was investigated (from 1.3 to 1.8 m^2). Dialysis was simulated using two different eluents: reverse osmotic water and 17.2% ethanol [32].

In agreement with the study of Haishima et al., the highest levels were measured when 17.2% ethanol was used, ranging from 54.8 to 4299 ng/dialyzer. Using osmotic water as eluent, the BPA levels measured span from 6.4 to 71.3 ng/dialyzer.

Additional factors influence the BPA released from dialyzers, like the type of dialyzer, surface area, and duration of dialysis session. Generally, large amount of BPA is released in long dialysis sessions or in dialyzer with high surface area.

The contribution of the PVC tubing to total BPA content in the eluates was negligible, and the levels found were below the limit of quantification.

Krieter et al. [33] also reported release of BPA from three different high and low flux dialyzers with different surface area (from 1.3 to 1.7 m^2) and with polycarbonate housing. BPA-free sterile water was circulated through the blood and dialysate compartments at 37°C, and BPA was measured by ELISA method. The amount of BPA eluted was significantly different between dialyzers evaluated, with average levels from 140.8 ± 38.7 to 6.2 ± 2.5 ng/dialyzer. These results are in the range with those reported in other studies when using water as eluent. The highest BPA levels were eluted from the low-flux dialyzer with polysulfone membrane, and the lowest from the dialyzer with polyethersulfone (PES) membrane. A summary of data available from the literature is presented in **Table 2**.

Table 2 summarizes the levels of BPA released by dialyzers and measured in different fluids.

Reference	Sample	Type of fluid (units)	BPA concentration
Haishima et al. [30]	PC housing/PS fibers	Water (A) Bovine serum (B) (ng/module)	(A): 31.0/141.8 (B): 1010/2090
	PC housing/CA fiber		(A): 34.1 (B): 196.1
	Polystyrene housing/PS fiber		(A): 3.78 (B): 140.7
Murakami et al. [31]	(A) PS fiber (B) PEPA fiber	Extraction with hexane (ng/mg fiber)	(A): 8.33 (B): 12.55
Fink [32]	Dialyzers with PS or PC (1.3–1.8 m^2)	(A) Osmotic water (B) 17.2 % EtOH (ng/dialyzer)	(A): 6.4–71.3 (B): 54.8–4299
Krieter et al. [33]	1.3 PS HF fiber (PC housing)	Water recirculation (400 ml), 3 h, 250 ml/min; 37°C (ng/dialyzer)	48.1 ± 7.7
	1.3 PS LF fiber (PC housing)		140.8 ± 38.7
	1.7 PES HF fiber (PC housing)		6.2 ± 2.5

Table 2. Levels of BPA released by dialyzers and measured in different fluids reported in the literature.

5. BPA in chronic kidney disease

The large number of molecules that accumulate in chronic kidney disease (CKD) are responsible for the uremic symptoms and contribute to increasing comorbidities and mortality in patients undergoing extracorporeal blood purification.

Removal of uremic toxins then is accompanied by an improvement in the clinical situation. They have been classified from the EuTox in differ groups according to their size and molecular weight [34]. A first group of molecules, around 350 toxins, is composed of small uremic toxins with molecular weight below 500 Da. Another group is characterized by medium-size toxins with molecular weight between 500 and 5000 Da. A new, and important, group or uremic toxins are represented by molecules with high affinity for proteins (protein-bound uremic toxins), which hamper their clearance.

One of the best characterized class of protein-bound toxins is phenols and indoles, a metabolite of protein catabolism by intestinal bacteria that have been related to renal failure progression and vascular damage in CKD. Proteins and peptides coming from diet are degraded by proteases and peptidases to simple amino acids. Some part of those amino acids reach the colon and are degraded by intestinal bacteria. This degradation generates potentially toxic metabolites such as ammonium, amines, thiols, phenols, and indoles.

Bisphenol A contains phenolic rings with structural similarity to phenols. While the origin of the toxins differs, the metabolism and side effects of BPA may have common characteristics with phenols of intestinal origin. BPA is eliminated by the kidney, and increased blood levels have been observed in CKD. Like intestinal phenols, after glucuronization in the liver, BPA is rapidly eliminated by the kidneys with a half-life in blood of less than 2 hours after oral ingestion that generally results in low blood levels. On the contrary, patients with impaired renal function could have BPA accumulation due to the less urinary excretion.

The National Health and Nutrition Examination Survey 2003–2006 (NHANES III) observed in 2573 patients a decrease of urinary excretion of BPA with renal function impairment [35]. The meaning of these data is uncertain: low urinary BPA excretion may reflect low exposure to BPA (which would be desirable) or retention of BPA by kidney disease (which would not be desirable).

By contrast, increased serum BPA with decreasing renal function and higher levels in hemodialysis was observed in different studies, suggesting that BPA may accumulate in CKD [31–33].

Additionally, the fractions of protein-bound and free plasma BPA in the maintenance of dialysis patients are 74 ± 5 and 26 ± 5%, respectively [33]. Moreover, a tissue/blood partition coefficient of 1.4 for nonadipose tissues and 3.3 for fat tissues further compromises the dialyzability of BPA [36], implying a concentration gradient highly in favor of driving BPA from dialysate to patient's blood.

Besides the other environmental sources, patients with end-stage renal disease on hemodialysis are repeatedly exposed to BPA from components of the dialyzer, more specifically polycarbonate housings and some dialysis membranes, resulting in a higher BPA blood levels of ESRD patients than healthy subjects.

BPA is found in the housing (polycarbonate) and membranes of some commonly used dialyzers; in **Table 3**, a summary of BPA contents in different dialyzers available on the market is provided.

Dialyzer BPA content is variable and depends on the manufacturer. All housings made with polycarbonate contain BPA (as a starting material), while the BPA contents in the fibers are variable. Generally, polysulfone membrane contains BPA in different amounts depending on the dialyzer surface. Other membranes such as polyethersulfone, polyarylethersulfone, polyamide, and polymethyl methacrylate are "naturally" BPA free.

The amount of BPA released depends on the experimental conditions and is higher in dialyzers perfused with blood than when perfused with saline. This difference has been attributed to the effect of blood hydrophobic components such as lipids or lipoproteins to extract BPA from the medical devices. An additional source of uncertainty on the quantitative determination of BPA is the analytical method used for determination. The most sensitive analytical method is represented by HPLC coupled with mass spectroscopy. Recently, simple ELISA methods are available on the market for BPA determination in biological fluids.

BPA may leach into aqueous solutions due to hydrolysis of the ester bonds in BPA-based polymers. With respect to the clinical situation, blood and water may differ in their ability to leach BPA. However, Krieter et al. [33] showed that considerable amounts of BPA were eluted from dialyzers with polysulfone membranes in 180 min of simulating (*in vitro*) dialysis conditions

http://dx.doi.org/10.5772/intechopen.68681

Company	Model name	BPA content		Company	Model name	BPA content	
		Fiber	Housing			Fiber	Housing
ALLMED	Polypure	Yes	Yes	FRESENIUS	Optiflux	Yes	Yes
ASAHI KASEI	Rexeed	Yes	No	FRESENIUS	F S	Yes	Yes
ASAHI KASEI	Leoceed	Yes	No	FRESENIUS	F HPS	Yes	Yes
ASAHI KASEI	APS	Yes	No	FRESENIUS	F	Yes	Yes
ASAHI KASEI	ViE	Yes	No	GAMBRO	Polyflux Revaclear	No	Yes
BAIN	BNoH	No	Yes	GAMBRO	Polyflux	No	Yes
BAXTER	Xenium XPH	No	Yes	GAMBRO	Evodial	No	Yes
BAXTER	Xenium	No	Yes	KAWASUMI	RENAK PS	Yes	Yes
BAXTER	Xenium PLUS	No	No	MEDICA	Smartflux	No	Yes
BAXTER	Exeltra	No	Yes	NIKKISO	FDY	No	Yes
BAXTER	Dicea	No	Yes	NIKKISO	FDX	No	Yes
BBRAUN	Xevonta	Yes	Yes	NIPRO	Solacea	No	No
BBRAUN	Diacap	Yes	Yes	NIPRO	Elisio	No	No
BELLCO	Phylther	No	Yes	NIPRO	Pureflex	No	No
BELLCO	Phylther UP	No	No	NIPRO	Sureflux	No	No
BELLCO	BLS	No	Yes	SERUMWERK	VITAPES	No	Yes
BELLCO	BLS UP	No	No	TORAY	TS	Yes	Yes
FRESENIUS	FX cordiax	Yes	No	TORAY	CS	Yes	No
FRESENIUS	FX	Yes	No	TORAY	Filtryzer	No	No

Table 3. Summary of BPA contents in different dialyzers available on the market.

with sterile water at body temperature. Previous studies had already shown that different polysulfone membranes leach varying amounts of BPA [31–37]. Obviously, this is also true for polysulfone membranes from the same manufacturer, indicating variations in different polysulfone lots or different extraction processes during fiber spinning. Additionally, a different amount of BPA is eluted from low-flux polysulfone membranes compared with high-flux polysulfone membranes. This difference, higher in HF membranes, may be attributed to a higher polymer content; usually less permeable low-flux membranes have a tighter wall structure compared with high-flux membranes. Since BPA is not a starting material of polyethersulfone membranes, the very small amounts of BPA eluted from this dialyzer most likely originate, in some cases, from the polycarbonate housing. Moreover, Krieter in his study also found that no differences in BPA levels were determined between the blood and dialysate compartments. This was because the unbound BPA can easily pass the dialysis membrane and equilibrate during recirculation. In vitro experiment suggests that dialyzers differing in elutable BPA used during chronic hemodialysis would have an impact on plasma BPA levels.

Recently, Bosch-Panadero et al. [38] performed a cross-over study to evaluate the impact of the dialyzer choice (BPA-free versus BPA-containing) on serum and intracellular BPA levels and on inflammation and oxidative stress markers in a group of 69 prevalent patients on hemodialysis.

The main finding of this study was that the choice of dialyzer in terms of BPA content impacts on acute (after a single dialysis session) and chronic (after 3 months of continuous use of the same type of dialyzer) changes in serum BPA levels. This reinforces the hypothesis of Krieter et al. that dialyzer BPA content may contribute to BPA burden in patients on hemodialysis.

The expression of oxidative stress markers was significantly higher after 3 months of hemodialysis with BPA-containing membranes with respect to BPA-free dialyzers. Three months of hemodialysis with BPA-containing membranes increased significantly circulating C-reactive protein (CRP) and IL-6 with respect to BPA-free dialyzers. These patients are more sensitive to BPA accumulation and potential toxicity due to the loss of the physiologic BPA excretion mechanisms in urine. In this same work, authors indicated that the serum BPA levels were 35-fold higher in patients on hemodialysis than in healthy controls confirming that serum BPA levels increased with decreasing renal function and are highest in individuals on hemodialysis.

A particular group of hemodialysis patients are those with diabetes. Recently, in a cross-over study, values of serum BPA have been measured in a group of 47 patients in which 12 had diabetes [39]. All patients were treated with low-flux polysulfone dialyzers.

In this study, postdialysis serum levels of BPA were significantly higher than predialysis levels. Additionally, diabetic patients showed higher predialytic BPA levels compared with those without diabetes. This difference disappeared for postdialysis measurement.

Unfortunately, no association was found between serum BPA levels and age, body mass index, dialytic vintage, blood pressure, and other medical parameters, probably due to the small number of subject investigated.

Up to now, we analyzed the dialyzer contribution to BPA in the blood of hemodialyzed patients, but recently, Bacle et al. [40] evaluated the potential exposure to BPA via the entire process of hemodialysis treatment, from production of purified water to dialysate and dialyzers. In their work, they could confirm that no BPA leaching is observed from bloodlines, confirming the information provided by the European PVC manufacturers who no longer use BPA in polyvinylchloride (PVC) production. At the same time, no leaching was observed from rinsing bags (0.9% sodium chloride), but larger amounts of BPA were found in dialyzers based on polysulfone and polycarbonate, as described by other authors and confirming the hypothesis that the dialyzers used during hemodialysis treatment may expose patients to a significant amount of BPA.

Concerning the water purification process, BPA has been detected in over 90% of collected samples, with significant amounts of BPA found after each step of the water treatment process. This suggests that none of the different processes applied in water purification is able to totally remove BPA. An additional source of BPA contamination already found in water of BPA could come from dialysis machine and dialysate cartridges, slightly increasing the BPA content in the dialysate.

In **Table 4**, a summary of selected study cited with the number of patients evaluated as well as BPA concentration in the serum samples is reported. Up to now, information provided by scientific literature on specific hemodialyzed population is scarce. Nevertheless, all studies reported showed an increase of BPA serum concentration in patients treated with BPA-containing dialyzers.

Reference	Number of patients	Type of study	Type of dialyzer (time of observation)	BPA concentration (ng/mL)
Turgut et al. [39]	47	Cross-over	PS (single session)	From 4.06 ± 0.73 to 5.57 ± 1.2
Bosch-Panadero et al. [38]	69	Cross over	PS (3 month)	From 48.8 ± 6.8 to 69.1 ± 10.1
			PN (3 month)	From 70.6 ± 8.4 to 47.1 ± 7.5
Krieter et al. [33]	18	Prospective	LF PS (4 weeks)	From 12.0 ± 6.0 to 10.52
			HF PS (4 weeks)	From 9.1 ± 4.5 to 8.72
			HF PES (4 weeks)	From 10.0 ± 4.9 to 8.45
Murakami et al. [31]	15	Cross-over	PS (3 month)	From 4.83 ± 1.94 to 6.62 ± 3.09
			Ce (1 month)	From 2.07 ± 2.10 to 1.48 ± 1.41
			PS (1 month)	From 3.78 ± 2.57 to 4.27 ± 2.98

Ce, cellulose; PS, polysulfone; PN, polynephron; PES, polyethersulfone; HF, high flux; LF, low flux.

Table 4. Levels of BPA measured in serum samples reported in the literature.

6. Conclusion

BPA is an estrogenic endocrine disruptor molecule with phenolic structure, used in the synthesis of polycarbonate plastics and epoxy resins. Exposition in the human population occurs mainly through the diet, in particular from food and beverages.

BPA could migrate into food from food and beverage containers with internal epoxy resin coatings and from products made of polycarbonate plastic such as tableware, food containers, and water bottles. BPA exposure results from either the release of unpolymerized monomers or the slow decay of polymer bonds in polycarbonate, leading to monomer release into foods and liquids. Starting from this information, data analysis coming from several large studies in various countries shows that the majority of the population examined have detectable levels of BPA conjugates in the urine. Indeed, in view of the rapid conjugation and elimination half-time of BPA, these levels reflect the exposure of the past hours just before the sample collection.

On the contrary, in patients with limited or absent kidney function, BPA may accumulate in the serum. The BPA accumulation in these subjects accounts from diet and medical device containing BPA, that is, extracted from the device by hydrophobic components present in the blood. Repeated loading of BPA during hemodialysis with BPA-containing membranes may aggravate the problem due to migration of BPA from dialyzers to the blood of patients and its inefficient removal due to the high protein-bound fraction of plasma BPA.

Some recent studies on the chronic use of BPA-free dialyzers indicate decrease of BPA serum levels in dialyzed patients reflecting a potential beneficial effect on inflammation and oxidative stress.

Furthermore, additional BPA contamination sources come from water and medical devices used to produce the dialysate fluid involved in hemodialysis treatment.

It is also advised that attention should be taken to avoid BPA cross contamination during medical devices production, with particular consideration to hemodialyzers. The possibility to replace BPA in these products should be assessed as well as the toxicological profile of the alternative materials. This issue could be a criterion for the purchase of medical devices commonly used in hemodialysis.

In conclusion, patients on hemodialysis have higher levels of serum and intracellular BPA with respect to healthy controls and the choice of dialysis membrane impacts on these levels. Dialyzers with BPA-containing membranes increase serum BPA levels. Studies indicate an increase of BPA serum concentration after a single dialysis session, confirming that hemodialysis does not compensate lack of urine BPA excretion.

Use of BPA-containing dialysis membranes further adds to the BPA burden of patients on hemodialysis. In contrast, it would be advisable the chronic use of BPA-free dialyzers to decrease BPA serum levels and related clinical effects.

Conflict of interest

The authors are full employees of Bellco (part of Medtronic) company. A company that produces and commercializes medical devices.

Author details

Giuseppe Palladino* and Luisa Sereni

*Address all correspondence to: giuseppe.palladino@medtronic.com

Scientific Affairs Department, Bellco, Mirandola, Italy

References

[1] Vogel SA. The politics of plastics: The making and unmaking of bisphenol A "Safety". American Journal of Public Health. 2009;**99**(Suppl 3):S559–S566. DOI: 10.2105/AJPH.2008.159228

[2] Rubin MM. Antenatal exposure to DES: Lessons learned…future concerns. Obstetrical & Gynecological Survey. 2007;**62**(8):548–555. DOI: 10.1097/01.ogx.0000271138.31234.d7

[3] Völkel W, Colnot T, Csanády GA, Filser JG, Dekant W. Metabolism and kinetics of bisphenol A in humans at low doses following oral administration. Chemical Research in Toxicology. 2002;**15**(10):1281–1287

[4] Calafat AM, Kuklenyik Z, Reidy JA, Caudill SP, Ekong J, Needham LL. Urinary concentrations of bisphenol A and 4-nonylphenol in a human reference population. Environmental Health Perspectives. 2005;**113**(4):391–395

[5] Calafat AM, Ye X, Wong LY, Reidy JA, Needham LL. Exposure of the US population to bisphenol A and 4-tertiary-octylphenol: 2003–2004. Environmental Health Perspectives. 2008;**116**(1):39–44. DOI: 10.1289/ehp.10753

[6] Bushnik T, Haines D, Levallois P, Levesque J, Van Oostdam J, Viau C. Lead and bisphenol A concentrations in the Canadian population. Health Reports. 2010;**21**(3):7–18

[7] Zhang J, Cooke GM, Curran IH, Goodyer CG, Cao XL. GC-MS analysis of bisphenol A in human placental and fetal liver samples.. Journal of Chromatography B, Analytical Technologies in the Biomedical and Life Sciences. 2011;**879**(2):209-214. DOI: 10.1016/j.jchromb.2010.11.031

[8] Mose T, Mathiesen L, Karttunen V, Nielsen JK, Sieppi E, Kummu M, et al. Meta-analysis of data from human ex vivo placental perfusion studies on genotoxic and immunotoxic agents within the integrated European project NewGeneris. Placenta. 2012;**33**(5):433–439. DOI: 10.1016/j.placenta.2012.02.004

[9] Balakrishnan B, Henare K, Thorstensen EB, Ponnampalam AP, Mitchell MD. Transfer of bisphenol A across the human placenta. American Journal of Obstetrics and Gynecology. 2010;**202**(4):e1-e7. DOI: 10.1016/j.ajog.2010.01.025

[10] Lang IA, Galloway TS, Scarlett A, Henley WE, Depledge M, Wallace RB, et al. Association of urinary bisphenol A concentration with medical disorders and laboratory abnormali-

ties in adults. Journal of the American Medical Association. 2008;**300**(11):1303–1310. DOI: 10.1001/jama.300.11.1303

[11] Melzer D, Rice NE, Lewis C, Henley WE, Galloway TS. Association of urinary bisphenol a concentration with heart disease: Evidence from NHANES 2003/06. PLoS One. 2010;**5**(1):e8673. DOI: 10.1371/journal.pone.0008673

[12] Melzer D, Harries L, Cipelli R, Henley W, Money C, McCormack P, et al. Bisphenol A exposure is associated with in vivo estrogenic gene expression in adults. Environmental Health Perspectives. 2011;**119**(12):1788–1793. DOI: 10.1289/ehp.1103809

[13] Carwile JL, Michels KB. Urinary bisphenol A and obesity: NHANES 2003–2006. Environmental Research. 2011;**111**(6):825–830. DOI: 10.1016/j.envres.2011.05.014

[14] Silver MK, O'Neill MS, Sowers MR, Park SK. Urinary bisphenol A and type-2 diabetes in US adults: Data from NHANES 2003-2008. PLoS One. 2011;**6**(10):e26868. DOI: 10.1371/journal.pone.002686

[15] Shankar A, Teppala S, Sabanayagam C. Urinary bisphenol A levels and measures of obesity: Results from the National Health and Nutrition Examination survey 2003–2008. ISRN Endocrinology. 2012;**2012**:965243. DOI: 10.5402/2012/965243

[16] Melzer D, Osborne NJ, Henley WE, Cipelli R, Young A, Money C, et al. Urinary bisphenol A concentration and risk of future coronary artery disease in apparently healthy men and women. Circulation. 2012;**125**(12):1482–1490. DOI: 10.1161/CIRCULATIONAHA.111.069153.

[17] Wang T, Li M, Chen B, Xu M, Xu Y, Huang Y, et al. Urinary bisphenol A (BPA) concentration associates with obesity and insulin resistance. The Journal of Clinical Endocrinology and Metabolism. 2012;**97**(2):E223–E227. DOI: 10.1210/jc.2011-1989

[18] Clayton EM, Todd M, Dowd JB, Aiello AE. The impact of bisphenol A and triclosan on immune parameters in the US Population, NHANES 2003–2006. Environmental Health Perspectives. 2011;**119**:390–396. DOI: 10.1289/ehp.1002883

[19] Miyawaki J, Sakayama K, Kato H, Yamamoto H, Masuno H. Perinatal and postnatal exposure to bisphenol A increases adipose tissue mass and serum cholesterol level in mice. Journal of Atherosclerosis and Thrombosis. 2007;**14**(5):245–252. DOI: 10.5551/jat.E486

[20] Somm E, Schwitzgebel VM, Toulotte A, Cederroth CR, Combescure C, Nef S, et al. Perinatal exposure to bisphenol A alters early adipogenesis in the rat. Environmental Health Perspectives. 2009;**117**(10):1549–1555. DOI: 10.1289/ehp.11342

[21] Alonso-Magdalena P, Morimoto S, Ripoll C, Fuentes E, Nadal A. The estrogenic effect of bisphenol A disrupts pancreatic beta-cell function in vivo and induces insulin resistance. Environmental Health Perspectives. 2006;**114**(1):106–112

[22] Alonso-Magdalena P, Vieira E, Soriano S, Menes L, Burks D, Quesada I, et al. Bisphenol A exposure during pregnancy disrupts glucose homeostasis in mothers and adult male offspring. Environmental Health Perspectives. 2010;**118**(9):1243–1250. DOI: 10.1289/ehp.1001993

[23] Nadal A. Obesity: Fat from plastics? Linking bisphenol A exposure and obesity. Nature Reviews. Endocrinology. 2013;**9**(1):9–10. DOI: 10.1038/nrendo.2012.205

[24] Ryan KK, Haller AM, Sorrell JE, Woods SC, Jandacek RJ, Seeley RJ. Perinatal exposure to bisphenol A and the development of metabolic syndrome in CD-1 mice. Endocrinology. 2010;**151**(16):2603-2612. DOI: 10.1210/en.2009–1218

[25] Wei J, Lin Y, Li Y, Ying C, Chen J, Song L, et al. Perinatal exposure to bisphenol A at reference dose predisposes offspring to metabolic syndrome in adult rats on a high-fat diet. Endocrinology. 2011;**152**(8):3049–3061. DOI: 10.1210/en.2011-0045

[26] Mackay H, Patterson ZR, Khazall R, Patel S, Tsirlin D, Abizaid A. Organizational effects of perinatal exposure to bisphenol-A and diethylstilbestrol on arcuate nucleus circuitry controlling food intake and energy expenditure in male and female CD-1 mice. Endocrinology. 2013;**154**(4):1465–1475. DOI: 10.1210/en.2012-2044

[27] Angle BM, Do RP, Ponzi D, Stahlhut RW, Drury BE, Nagel SC, et al. Metabolic disruption in male mice due to fetal exposure to low but not high doses of bisphenol A (BPA): Evidence for effects on body weight, food intake, adipocytes, leptin, adiponectin, insulin and glucose regulation. Reproductive Toxicology. 2013;**42**:256–268. DOI: 10.1016/j.reprotox.2013.07.017

[28] Yan S, Chen Y, Dong M, Song W, Belcher SM, Wang HS. Bisphenol A and 17β-estradiol promote arrhythmia in the female heart via alteration of calcium handling. PLoS One. 2011;**6**(9):e25455. DOI: 10.1371/journal.pone.0025455

[29] Testai E, Hartemann P, Rodríguez-Farre E, Rastogi SC, Bustos J, Gundert-Remy U, et al. The safety of the use of bisphenol A in medical devices. Regulatory Toxicology and Pharmacology. 2016;**79**:106–107. DOI: 10.1016/j.yrtph.2016.01.014

[30] Haishima Y, Hayashi Y, Yagami T, Nakamura A. Elution of bisphenol-A from hemodialyzers consisting of polycarbonate and polysulfone resins. Journal of Biomedical Materials Research. 2001;**58**(2):209–2015

[31] Murakami K, Ohashi A, Hori H, Hibiya M, Shoji Y, Kunisaki M, et al. Accumulation of bisphenol A in hemodialysis patients. Blood Purification. 2007;**15**(3):290–294. DOI: 10.1159/000104869

[32] Fink K. Toxins in Renal Disease and Dialysis Therapy: [dissertation]. Bayerischen Julius-Maximilians-Universität Würzburg:2008. 167 p.

[33] Krieter DH, Canaud B, Lemke HD, Rodriguez A, Morgenroth A, von Appen K, et al. Bisphenol A in chronic kidney disease. Artificial Organs. 2013;**37**(3):283–290. DOI: 10.1111/j.1525-1594.2012.01556.x

[34] Vanholder R, Baurmeister U, Brunet P, Cohen G, Glorieux G, Jankowski J, et al. A bench to bedside view of uremic toxins. Journal of the American Society of Nephrology: JASN. 2008;**19**(5):863–870. DOI: 10.1681/ASN.2007121377

[35] You L, Zhu X, Shrubsole MJ, Fan H, Chen J, Dong J, et al. Renal function, bisphenol A, and alkylphenols: Results from the National Health and Nutrition Examination Survey (NHANES 2003-2006). Environmental Health Perspectives. 2011;**119**(4):527–533. DOI: 10.1289/ehp.1002572

[36] Csanády GA, Oberste-Frielinghaus HR, Semder B, Baur C, Schneider KT, Filser JG. Distribution and unspecific protein binding of the xenoestrogens bisphenol A and daidzein. Archives of Toxicology. 2002;**76**(5-6):299–305. DOI: 10.1007/s00204-002-0339-5

[37] Yamasaki H, Nagake Y, Makino H. Determination of bisphenol A in effluents of hemodialyzers. Nephron. 2001;**88**(4):376–378. DOI: 46023

[38] Bosch-Panadero E, Mas S, Sanchez-Ospina D, Camarero V, Pérez-Gómez MV, Saez-Calero I, et al. The choice of hemodialysis membrane affects bisphenol A levels in blood. Journal of the American Society of Nephrology: JASN. 2016;**27**(5):1566–1574. DOI: 10.1681/ASN.2015030312

[39] Turgut F, Sungur S, Okur R, Yaprak M, Ozsan M, Ustun I et al. Higher serum bisphenol A levels in diabetic hemodialysis patients. Blood Purification. 2016;**42**(1):77–82. DOI: 10.1159/000445203

[40] Bacle A, Thevenot S, Grignon C, Belmouaz M, Bauwens M, Teychene B et al. Determination of bisphenol A in water and the medical devices used in hemodialysis treatment. International Journal of Pharmaceutics. 2016;**505**(1-2):115–121. DOI: 10.1016/j.ijpharm.2016.03.003

Chapter 5

Toxicogenomics of Bisphenol A and Neurodevelopmental Disorders

Bingling Wang, Ruqin Gao and Da-Hong Wang

Additional information is available at the end of the chapter

http://dx.doi.org/10.5772/intechopen.68415

Abstract

Bisphenol A (BPA) has been widely used in many industrial and consumer products and is known as an endocrine-disrupting chemical. To find the underlying genetic basis and molecular mechanisms of BPA-associated neurodevelopmental disorders (NDs), this chapter addressed the toxicogenomics of BPA with publicly accessed Comparative Toxicogenomics Database. The present results indicated that the key cellular components (CC) of the nervous system such as neuron, synapse, dendrite and axon are common in CC annotation; the commonly found molecular functions are neurotransmitter receptor or transducer binding or activity; and the main common biological processes include synaptic signalling, cognition, learning or memory, behaviour, the development of nervous system and brain. Neuroactive ligand-receptor interaction, dopaminergic, glutamatergic and serotonergic synapses, monoamine transport and synaptic vesicle pathway were the common pathways. Simultaneously, the BPA-disease may share the common pathways with drug addictions such as cocaine addiction. Unique pathways might also contribute to the BPA action in different NDs such as one carbon metabolism and detoxification of oxidative stress in Down syndrome. Although GO and pathway results indicate some common annotations, the predicted PPI molecular function clusters are quite different for each ND. In addition, some of the NDs share the same transcription factors (TFs) and miRNAs, which indicate these disorders have the similar expression profiles. Finally, chemicals having comparable interacting genes to BPA should be considered.

Keywords: bisphenol A, toxicogenomics, neurodevelopmental disorders

1. Introduction

The critical windows of vulnerability during human brain development are mainly from the third trimester to at least 2–3 years after birth [1]. Any developmental neurotoxicant exposure during this critical window has the possibility to cause various clinical neurodevelopmental disorders (NDs) in humans (e.g. autism, anxiety disorder, schizophrenia, dyslexia and epilepsy). Bisphenol A (BPA), as a proved endocrine-disrupting chemical, has been widely used as the plasticizer in many consumer products made of polycarbonate plastic such as baby bottles, tableware, food containers and water bottles. BPA can also be found in breast milk [2, 3]. Infants and children were found to have the highest estimated daily intake of BPA per body weight [4, 5]. A review by Healy et al. [6] affirmed that the potential for non-dietary sources make a substantial contribution to the total daily BPA exposure in young children and recommended risk-assessment models implement new frameworks, which specifically address exposure and hazard in early childhood. Pinson et al. [7] reviewed the human and rodent data on the neurodevelopmental alterations of BPA, and found that mostly reported effects were social and sexual behaviour and cognition that were unique to humans. The related mechanisms reported included the disruption of thyroid function, alterations of neurotransmitters levels, calcium signalling and neurotoxicity. Given the extensive BPA exposure during the critical windows of the brain development and the possible neurodevelopmental alterations, here we explore the possible genetic basis and the molecular mechanisms of BPA-associated NDs.

2. BPA-gene interactions and neurodevelopmental disorders

Comparative toxicogenomics database (CTD, http://ctdbase.org) is a robust, publicly available database that aims to advance understanding about how environmental exposures affect human health with manually curated information about chemical-gene/protein interactions, chemical-disease and gene-disease relationships, with functional and pathway data to aid in the development of hypotheses about the mechanisms underlying environmentally influenced diseases [8]. In this work, all BPA-gene/protein interactions were downloaded from CTD, in which BPA-gene/protein interactions associated to the following 17 NDs were selected as our targeted NDs for further analysis according to MESH ID used in CTD—anxiety disorders (AD), attention deficit and disruptive behaviour disorders (ADDBD), autism spectrum disorder (ASD), bipolar disorder (BD), developmental disabilities (DD), Down syndrome (DS), foetal alcohol spectrum disorders (FASD), intellectual disability (ID), language development disorders (LDD), learning disorders (LD), motor skills disorders (MSD), obsessive-compulsive disorder (OCD), pervasive child development disorders (PCDD), schizophrenia (Sch), speech disorders (SD), stereotypic movement disorder (SMD) and Tourette syndrome (TS). Thus, BPA-gene/protein interactions associated to these 17 NDs were collected for further analysis.

According to the reference score on relationships between chemicals-genes, genes-diseases and chemicals-diseases [9], we found that ID was most likely having the atypical connectivity with BPA (**Table 1**). Inference BPA interacted genes were up to 119. Inference score of more

Disease name	Inference BPA-interacted genes (n)	Inference score	Reference count
ID	ACBD6, ADK, ADRA2B, AHI1, ALDH5A1, AP4E1, AP4M1, APC, ARL14EP, ASCC3, ASCL1, BBS7, BDNF, CA8, CACNA1G, CALCA, CAPN10, CASP2, CCBE1, CCNA2, CIC, CNDP1, CNKSR1, COL18A1, DEAF1, DISC1, DNMT3A, DOCK8, DYNC1H1, EEF1B2, ELP2, ENTPD1, ERLIN2, FAM126A, FASN, FGFR2, FMR1, FOLR1, FRY, GAMT, GNAS, GON4L, GRIN2B, HDAC4, HEXA, HIST3H3, INPP4A, INPP5E, KCNA2, KDM5A, KDM5C, KDM6B, KIF1BP, KIF7, L2HGDH, LAMA1, LARP7, LETM1, LINS1, MAN1B1, MCC, MECP2, MED13L, MEF2C, METTL23, MFSD2A, NAGLU, NDST1, NF1, NRXN1, NSD1, PARP1, PAX6, PDHX, PECR, PEX6, PMM2, POLR3B, PRKCG, PRKRA, PTCHD1, PTEN, RAB39B, RABL6, RALGDS, RGS7, SC5D, SCAPER, SCN8A, SETBP1, SHANK2, SHANK3, SIN3A, SLC2A1, SLC31A1, SLC4A10, SNX14, SRD5A3, SRGAP3, STRA6, SURF1, SYNGAP1, TAF2, TH, TMCO1, TMEM135, TRMT1, TSEN2, TSEN34, TSEN54, TTI2, UBR7, UROC1, VIP, VRK1, WDR45B, WDR62, ZBTB40, ZCCHC8 (119)	86.31	51
LD	ACHE, APOD, APP, BCL2, CAMKMT, GRIA1, HMOX1, HTR1A, HTR7, IL1B, IL1RN, KL, MAPT, MECP2, MICU1, MT1, NF1, PARK2, PDE1B, PNOC, POR, PSEN1, RNF135, SIGMAR1, SLC17A6, SLC17A7, SYP, TH, TRH, VEGFA (30)	30.96	39
Sch	ACOT6, ADAMTS3, ADCY7, ADGRF4, AHI1, AKT1, ALS2CL, APOE, AVP, BDNF, BTG1, CAMK2B, CASP4, CCDC137, CCL2, CELF2, CFAP65, CHD4, CHI3L1, CLINT1, CNR1, COL3A1, COMT, CP, CPLX1, CPLX2, DAO, DGCR2, DISC1, DIXDC1, DLG1, DPYD, DRD1, DRD2, DRD3, DRD4, DTNBP1, EDEM2, EIF5, ESAM, FAM3D, FASTKD5, FGFR1, GABRA6, GABRB2, GABRD, GAD1, GAD2, GIF, GPR153, GRIK2, GRIK5, GRIN2B, GRIN2D, GRM2, GRM3, GSK3A, GSK3B, HCAR2, HLA-DRB1, HNRNPA3, HP, HRH1, HTR2A, HTR6, HTR7, IL1B, IL2RA, IL6, IL6R, INPP5A, KDM2B, KDR, KLF12, KPNA1, LAMA1, LAMA2, LGR4, LRP1, MAGI2, MAOB, MET, MTHFR, MTOR, ND4, NDUFV2, NKAPL, NOS1, NPRL2, NR3C1, NRG1, NRG3, NRGN, NRIP1, NRXN1, NTF3, NTNG1, NTNG2, NTRK1, NTSR1, OXTR, PAK2, PCM1, PDE4B, PHB, PIK3CB, PITPNM1, PLCB1, PLCL2, PLXNA2, PML, PRODH, PVALB, RB1CC1, RELN, RGS12, RGS4, RGS9, RTN4, RTN4R, SAP30BP, SBNO1, SDF4, SELENBP1, SLC26A7, SLC6A1, SLC6A3, SLC6A4, SP4, SPATA5, SRSF1, SYN2, SYP, TAAR6, TAC1, TEKT5, THBS1, TNF, TP53, TPH1, TRAK1, TRRAP, TSPAN18, UGT1A3, VIPR2, VPS35, VPS39, WDR11, ZKSCAN4, ZNF565 (150)	30.91	86
ASD	AVPR1A, BDNF, C3ORF58, CEP41, CHD8, CIRBP, CXORF36, DHCR24, DIO2, DIO3, DLG4, DLX1, DNMT3A, DNMT3B, DPP6, DPYD, EN2, FOXP1, GABRB3, GRIN2B, GTF2I, HEY1, HFE, IL1RAPL1, ITGB3, JARID2, LAMC3, LRRN3, LRRTM3, MEF2C, MTNR1A, NRXN1, NRXN2, NTSR1, OXTR, PCDH9, PTCHD1, RELN, RYR2, SCN1A, SFSWAP, SHANK3, SIN3A, SNTG2, SOX5, SOX9, TBL1X, TET1, TET3, TSHZ3, UPP2 (51)	24.05	29
AD	ADORA2A, APP, CARTPT, CHRNA5, CHRNB2, CNR1, CRH, CRHR2, CRP, DIXDC1, DNMT1, DRD2, EOMES, FOS, GABRA2, GLO1, GNB1, GRM8, HTR7, MAGI2, MDK, MECP2, MIF, NPS, NPY, NPY1R, OXT, PAM, SERPINA1, SHANK1, SLC6A3, SLC6A4, TNF, UCN (34)	21.95	52
SMD	CRH, MEF2C, TRH (3)	13.04	3
BD	AKR1C4, ANK3, BDNF, BHLHE40, CACNA1C, COMT, CPLX1, CPLX2, DIXDC1, DRD1, DRD5, GRIK2, GRK3, GSK3A, GSK3B, HTR2A, INS, MTHFR, NDUFV2, NR3C1, NTNG1, NTNG2, NTRK1, NTRK2, PDE4B, POMC, PVALB, RELN, S100B, SERPINA1, SLC5A3, SLC6A4, SNAP25, SP4, TAC1, TACR1, TENM4, TRPC3, TSHB (39)	12.32	31

Disease name	Inference BPA-interacted genes (n)	Inference score	Reference count
SD	GRIN2A, MFSD2A, TTPA (3)	12.24	3
DS	CALCA, CXCL8, GATA1, GSTM2, MTHFR, MTR, NTF3, PRDX2, PRDX6, RCAN1, S100B, SLC19A1, SOD1, VIP (14)	9.35	10
TS	DRD3, SLITRK1 (2)	8.31	2
MSD	AKAP5, CAMKMT, FGFR2, OGG1, PTEN, SHANK1 (6)	7.05	6
OCD	BDNF, CCKBR, HOXB8, HTR1D, HTR2A, SLC6A4, SLITRK5 (7)	5.95	7
ADDBD	DRD4, S100B (2)	4.03	2
LDD	BCL11A, DPYD, ERF, FOXP2, GRIN2A, KCNA2, NRXN1, PTEN, SETBP1, SHANK3 (10)	3.52	10
DD	ARFGAP1, CAMKMT, CBL, CHRNA4, CNTN4, DOCK8, DRD2, KCNQ2, KCNT1, LRP2, MECP2, NANS, NTRK2, PMP22, PNKP, PTEN, SHANK3, SLC2A1, SLC33A1, SLC4A4, SLC6A8, STAMBP (22)	2.99	24
FASD	CAT, NOS1 (2)	2.49	2
PCDD	DRD4, MECP2, MKL2 (3)	2.47	3

ID: intellectual disability; LD: learning disorders; Sch: schizophrenia; ASD: autism spectrum disorder; AD: anxiety disorders; SMD: stereotypic movement disorder; BD: bipolar disorder; SD: speech disorders; DS: Down syndrome; TS: Tourette syndrome; MSD: motor skills disorders; OCD: obsessive-compulsive disorder; ADDBD: attention deficit and disruptive behavior disorders; LDD: language development disorders; DD: developmental disabilities; FASD: foetal alcohol spectrum disorders; PCDD: pervasive child development disorders.

Table 1. Selected neurodevelopmental diseases and related BPA-interacted genes.

than 20 was found for LD, Sch, ASD and AD, whereas it was less than 10 for DS, TS, MSD, OCD, ADDBD, LDD, DD, FASD and PCDD. The results showed that it was only two inference BPA interacted genes for TS, ADDBD and FASD, and in total, 403 BPA bi-interacted genes were curated. A total of 563 BPA-mRNA bi-interactions were found, in which 240 expressions were down-regulated, 169 up-regulated and 153 were altered (not mentioned up or down) regulation. Simultaneously, eighty-one BPA-protein bi-interactions, two protein-BPA bi-interactions and eight BPA-DNA methylation interactions were reported.

3. Microarray data and differently expressed gene screening

To explore the possible clinical application of the genes curated, we used Gene Expression Omnibus (http://www.ncbi.nlm.nih.gov/geo/) and ArrayExpress (http: //www. ebi.ac.uk / arrayexpress) to find the microarray data for peripheral blood, and special tissue gene expression profiling in NDs. The differently expressed genes (DEGs) in each ND samples were identified. If more than one microarray dataset is found, all the datasets are integrated to perform the meta-analysis for the DEGs. Based on the inference scores and the counts of inference BPA-interacted genes, we only selected ID, LD, Sch, ASD, AD and BD as our target diseases for the possible microarray data.

We only found some common genes for BD between CTD and the microarray data. It was in the GSE46449 [10] and only four common genes (BDNF, CACNA1C, CPLX2, HTR2A, SP4) were found. Therefore, the existing microarray data is not enough for further annotation.

4. Gene function enrichment analysis

The curated genes in CTD for each ND were uploaded to DAVID 6.8 Beta (https://david-d.ncif-crf.gov/tools.jsp). Homo sapiens were used as the background population. The gene ontology (GO) and pathway were analysed [11, 12]. Simultaneously, WikiPathways and Reactome of EnrichR [13, 14] were used for duplicated pathway prediction. STRING (http://string-db.org) is a database of known and predicted protein-protein interactions (PPI), and can be used to predict the functional associations between proteins [15]. Based on the information provided by the STRING database, all genes related to each neurodevelopmental disorder are used to construct a PPI network. The functional molecules in the PPI network are subsequently identified by the molecular complex detection (MCODE) plugin [16] of Cytoscape [17]. The MCODE is a well-known automated method to find highly interconnected subgraphs as molecular complexes or clusters in a PPI network. The proteins in each module will be transferred in Genemania app [18] to predict the possible target proteins or biomarkers.

4.1. Intellectual disability

Of the 119 interacted genes, 117 were bi-interacted. GO analysis with these 117 genes indicated that BPA bi-interacted genes are involved in the biological processes (BP) such as cognition, the development of nervous system, head, brain, forebrain, pallium, telencephalon and embryonic organ (**Table 2**). Other BPs included learning or memory, behaviour (social, single- or multi-organism), intraspecies interaction between organisms, embryonic organ morphogenesis, regulation of synapse structure or activity and neuron apoptotic process. Of the BPs, nervous system development was also reported in a recent study on systematic phenomics analysis for the genes muted in ID by Kochinke et al. [19]. Some genes possibly involved in the cellular component (CC) of somatodendritic compartment and the molecular functions (MF) of chromatic binding. Pathway analysis only found MECP2 and associated Rett Syndrome in WikiPathways, in which six BPA bi-interacted genes were involved. This might suggest that ID and Rett syndrome possess the same pathway.

PPI analysis found four molecular modules (**Figure 1**). TSEN2, TSEN34, TSEN54 and VRK1 were involved in module 1. TSEN2, TSEN34 and TSEN54 are tRNA splicing endonuclease subunits and can interact through physical interaction or co-expression. VRK1 might interact with TSEN2, TSEN34 and TSEN54 through the same pathway, physical interaction or co-expression. CLP1 and TSEN15 can physically interact with TSEN2, TSEN34 and TSEN54 [20–22], therefore, BPA has the potential to interact with these two genes. BPA might also interact with WARS and WARS2 because of their predicted interaction with TSEN34 [23, 24]. WARS and WARS2 are involved in the tryptophan metabolic pathway, which have been reported to appear to provide a unifying biochemical basis for ASDs [25]. Tryptophan is a precursor of important compounds,

Term	Count	*P* value	FDR	Genes
BP				
Cognition	16	2.49E−10	4.53E−07	DEAF1, VIP, MEF2C, PTCHD1, TH, NF1, MECP2, PRKCG, NRXN1, PTEN, SHANK2, SHANK3, GRIN2B, GNAS, STRA6, SYNGAP1
Learning or memory	14	4.79E−09	8.72E−06	DEAF1, VIP, MEF2C, TH, NF1, MECP2, PRKCG, NRXN1, PTEN, SHANK2, SHANK3, GRIN2B, STRA6, SYNGAP1
Nervous system development	39	1.56E−08	2.85E−05	FGFR2, MEF2C, DEAF1, NAGLU, BBS7, NDST1, PTCHD1, KCNA2, TH, PAX6, MFSD2A, PTEN, BDNF, FOLR1, INPP5E, ADRA2B, CASP2, KIF1BP, DISC1, APC, DNMT3A, ALDH5A1, NF1, FMR1, MECP2, AHI1, PRKCG, NRXN1, SHANK2, SHANK3, FRY, ASCL1, HDAC4, SLC4A10, WDR62, SCN8A, SYNGAP1, KDM6B, FAM126A
Head development	22	2.28E−08	4.14E−05	MEF2C, FGFR2, NAGLU, BBS7, NDST1, PTCHD1, NF1, TH, PAX6, MECP2, AHI1, MFSD2A, NRXN1, PTEN, SHANK3, ASCL1, SLC4A10, WDR62, STRA6, CASP2, DISC1, KDM6B
Brain development	21	4.94E−08	0.0001	MEF2C, FGFR2, NAGLU, BBS7, NDST1, PTCHD1, NF1, TH, PAX6, MECP2, AHI1, MFSD2A, NRXN1, PTEN, SHANK3, ASCL1, SLC4A10, WDR62, CASP2, DISC1, KDM6B
Single-organism behaviour	16	9.96E−08	0.0002	DEAF1, VIP, MEF2C, KCNA2, TH, NF1, MECP2, PRKCG, NRXN1, PTEN, SHANK2, SHANK3, SLC4A10, GRIN2B, STRA6, SYNGAP1
Forebrain development	15	2.10E−07	0.0004	FGFR2, MEF2C, NDST1, PTCHD1, NF1, TH, PAX6, MFSD2A, PTEN, SHANK3, ASCL1, SLC4A10, WDR62, DISC1, KDM6B
Behaviour	18	2.56E−07	0.0005	DEAF1, VIP, MEF2C, NAGLU, PTCHD1, KCNA2, TH, NF1, MECP2, PRKCG, NRXN1, PTEN, SHANK2, SHANK3, SLC4A10, GRIN2B, STRA6, SYNGAP1
Central nervous system development	22	9.19E−07	0.0017	MEF2C, FGFR2, NAGLU, BBS7, NDST1, PTCHD1, ALDH5A1, NF1, TH, PAX6, MECP2, AHI1, MFSD2A, NRXN1, PTEN, SHANK3, ASCL1, SLC4A10, WDR62, CASP2, DISC1, KDM6B
Social behaviour	7	9.59E−07	0.0017	PTCHD1, TH, MECP2, NRXN1, PTEN, SHANK2, SHANK3
Intraspecies interaction between organisms	7	9.59E−07	0.0017	PTCHD1, TH, MECP2, NRXN1, PTEN, SHANK2, SHANK3
Pallium development	10	1.74E−06	0.0032	MEF2C, ASCL1, WDR62, NF1, TH, PAX6, MFSD2A, PTEN, KDM6B, DISC1
Embryonic organ morphogenesis	12	3.27E−06	0.0060	FGFR2, MEF2C, NAGLU, NDST1, BBS7, FOLR1, PRKRA, TH, PAX6, AHI1, STRA6, GNAS
Learning	9	4.59E−06	0.0084	DEAF1, NF1, TH, MECP2, STRA6, NRXN1, SYNGAP1, SHANK2, SHANK3

Term	Count	*P* value	FDR	Genes
Regulation of synapse structure or activity	11	4.63E−06	0.0084	MEF2C, BDNF, FMR1, NF1, MECP2, NRXN1, SYNGAP1, PTEN, SHANK2, SHANK3, DISC1
Multi-organism behaviour	7	5.03E−06	0.0092	PTCHD1, TH, MECP2, NRXN1, PTEN, SHANK2, SHANK3
Telencephalon development	11	5.17E−06	0.0094	MEF2C, ASCL1, WDR62, NF1, TH, PAX6, MFSD2A, PTEN, KDM6B, SHANK3, DISC1
Regulation of neuron apoptotic process	10	8.61E−06	0.0157	MEF2C, ASCL1, HDAC4, BDNF, NF1, MECP2, PRKCG, SYNGAP1, PARP1, CASP2
Neuron apoptotic process	10	1.27E−05	0.0231	MEF2C, ASCL1, HDAC4, BDNF, NF1, MECP2, PRKCG, SYNGAP1, PARP1, CASP2
Observational learning	4	1.58E−05	0.0287	NF1, STRA6, NRXN1, SHANK3
Embryonic organ development	13	2.24E−05	0.0408	MEF2C, FGFR2, NAGLU, BBS7, NDST1, FOLR1, PRKRA, TH, PAX6, AHI1, GAMT, STRA6, GNAS
CC				
Somatodendritic compartment	18	1.10E−05	0.0152	VIP, SNX14, KCNA2, FMR1, TH, NF1, PRKCG, NRXN1, PMM2, PTEN, SHANK2, SHANK3, ASCL1, SLC4A10, GNAS, SYNGAP1, SLC31A1, APC
MF				
Chromatin binding	15	1.23E−05	0.0180	TAF2, MEF2C, DNMT3A, FMR1, PAX6, MECP2, CIC, HDAC4, ASCL1, VRK1, SIN3A, KDM5A, HIST3H3, NSD1, KDM6B

Table 2. GO analysis for the genes related to intellectual disability.

such as serotonin, quinolinic acid and kynurenic acid, which are involved in neurodevelopment and synaptogenesis. Decreased tryptophan metabolism may alter brain development, neuro-immune activity and mitochondrial function. In module 2, ASCL1, HDAC4, BDNF, KDM6B, TH, HIST3H3, PAX6 and SIN3A interacted through co-localization, co-expression and physical interactions. TH and KDM6B linked other six node proteins through CALCOCO1 [26]. In module 3, LARP7, GAMT and TRMT1 were interacted by physical interactions, co-expression and predicted, pathway or genetic interaction. In this module, CDK9 plays an import role to link the three genes mainly by physical interactions [27–29], although there was no direct evidence for BPA-CDK9 interaction. CDK9 might associate with ID by JUN binding [30] or by AFF family of RNA-binding proteins [31]. In module 4, PARP1, FASN and PTEN interact through physical interaction, co-expression, predicted, pathway, genetic interaction and shared protein domains. Recent findings have proven that the mTOR pathway is altered in cells with defective DNA repair. PARP1 is related to the accumulation of irreparable DNA damage [32], while PTEN is a phosphatase to mediate switching off the PI3K/Akt/mTOR signalling pathway, which has been reportedly associated with ID [33–35]. FASN (expression of fatty acid synthase) is found negatively correlated with PTEN [36], but the in-between genes were not explored. FASN may co-express with MAST2 [26], PRKDC [26, 37] or BMI1 to physically interact with PTEN.

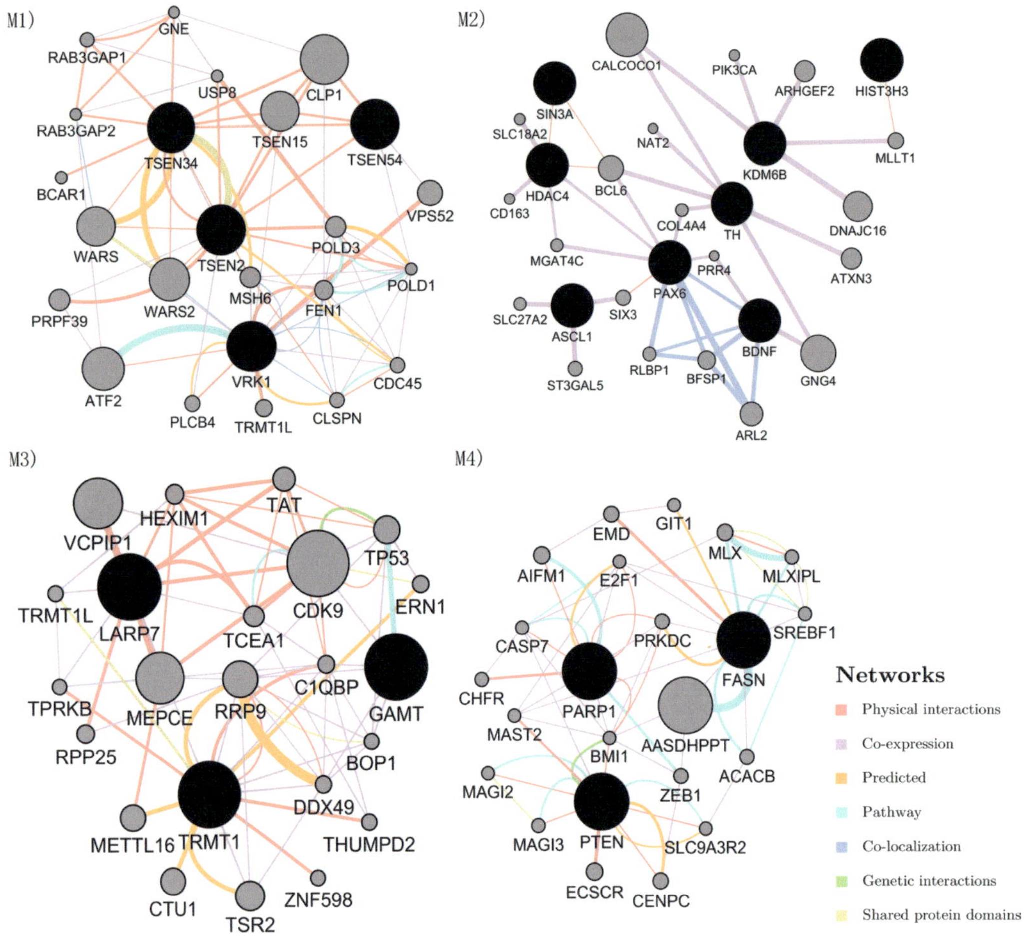

Figure 1. Networks for the genes in PPI MCODE molecular modules for ID.

4.2. Learning disorders

A total of 29 BPA bi-interacted genes related to LD were found. Some of these genes are involved in the BPs such as behaviour, learning or memory, cognition, synaptic-signalling related, cell-cell signalling related, secretion, neuron death or apoptotic-related processes (**Table 3**). Simultaneously, some of these genes may participate in the CCs such as synapse, presynapse, dendrite, somatodendritic compartment, axon or secretory vesicle. Five KEGG pathways might be influenced by BPA (**Table 4**). Retrograde endocannabinoid signalling, serotonergic synapse and glutamatergic synapse pathways may be influenced by BPA and involve in LD. Nicotine addiction and Alzheimer's disease may share the same pathway with LD. WikiPathways indicated monoamine transport, synaptic vesicle pathway, MECP2 and associated Rett syndrome pathway might also be influenced. Interestingly, some genes involved in LD are also found in sudden infant death syndrome (SIDS) susceptibility pathways. Reactome pathway confirmed the serotonergic synapse pathway found in KEGG, and additionally suggested that organic anion transporters pathway may be influenced by BPA.

Term	Count	*P* value	FDR	Genes
BP				
Single-organism behaviour	14	2.98E−14	5.29E−11	SLC17A7, APP, HTR1A, PSEN1, PDE1B, GRIA1, BCL2, MAPT, NF1, TH, MECP2, IL1B, PARK2, TRH
Behaviour	14	2.07E−12	3.67E−09	SLC17A7, APP, HTR1A, PSEN1, PDE1B, GRIA1, BCL2, MAPT, NF1, TH, MECP2, IL1B, PARK2, TRH
Learning or memory	10	9.30E−11	1.65E−07	SLC17A7, APP, PDE1B, PSEN1, GRIA1, NF1, TH, MECP2, IL1B, PARK2
Cognition	10	3.03E−10	5.39E−07	SLC17A7, APP, PDE1B, PSEN1, GRIA1, NF1, TH, MECP2, IL1B, PARK2
Chemical synaptic transmission	12	2.11E−09	3.75E−06	SYP, SLC17A7, ACHE, HTR1A, PNOC, PSEN1, GRIA1, HTR7, NF1, TH, MECP2, PARK2
Trans-synaptic signalling	12	2.11E−09	3.75E−06	SYP, SLC17A7, ACHE, HTR1A, PNOC, PSEN1, GRIA1, HTR7, NF1, TH, MECP2, PARK2
Synaptic signalling	12	2.11E−09	3.75E−06	SYP, SLC17A7, ACHE, HTR1A, PNOC, PSEN1, GRIA1, HTR7, NF1, TH, MECP2, PARK2
Anterograde trans-synaptic signalling	12	2.11E−09	3.75E−06	SYP, SLC17A7, ACHE, HTR1A, PNOC, PSEN1, GRIA1, HTR7, NF1, TH, MECP2, PARK2
Secretion by cell	13	2.26E−08	4.02E−05	SLC17A7, ACHE, APP, HTR1A, PSEN1, HMOX1, NF1, VEGFA, IL1RN, MECP2, IL1B, PARK2, TRH
Cell-cell signalling	15	3.65E−08	6.49E−05	ACHE, TH, IL1RN, NF1, MECP2, PARK2, TRH, SYP, SLC17A7, HTR1A, PNOC, PSEN1, GRIA1, HTR7, IL1B
Secretion	13	9.70E−08	1.73E−04	SLC17A7, ACHE, APP, HTR1A, PSEN1, HMOX1, NF1, VEGFA, IL1RN, MECP2, IL1B, PARK2, TRH
Regulation of cell communication	19	1.13E−07	2.01E−04	ACHE, KL, NF1, IL1RN, MECP2, PARK2, TRH, POR, SYP, APP, CAMKMT, HTR1A, APOD, PSEN1, GRIA1, BCL2, HMOX1, VEGFA, IL1B
Regulation of signalling	19	1.46E−07	2.60E−04	ACHE, KL, NF1, IL1RN, MECP2, PARK2, TRH, POR, SYP, APP, CAMKMT, HTR1A, APOD, PSEN1, GRIA1, BCL2, HMOX1, VEGFA, IL1B
System process	16	1.56E−07	2.78E−04	TH, NF1, MECP2, PARK2, SLC17A7, APP, HTR1A, PNOC, PSEN1, PDE1B, GRIA1, HTR7, BCL2, HMOX1, VEGFA, IL1B
Dicarboxylic acid transport	6	1.68E−07	2.98E−04	SLC17A7, SLC17A6, PSEN1, NF1, IL1B, TRH
Organic hydroxy compound metabolic process	9	2.69E−07	4.79E−04	APP, HTR1A, PDE1B, BCL2, TH, MECP2, IL1B, PARK2, POR
Neuron death	8	4.39E−07	7.81E−04	APP, PSEN1, HMOX1, BCL2, NF1, MECP2, PARK2, SIGMAR1
Locomotory behaviour	7	6.43E−07	1.14E−03	APP, PDE1B, MAPT, TH, MECP2, PARK2, TRH
Response to hypoxia	8	7.59E−07	1.35E−03	HMOX1, BCL2, VEGFA, NF1, TH, MECP2, IL1B, TRH

Term	Count	*P* value	FDR	Genes
Memory	6	7.93E−07	1.41E−03	SLC17A7, PSEN1, GRIA1, TH, MECP2, IL1B
Regulation of neurotransmitter levels	7	8.92E−07	1.59E−03	SLC17A7, ACHE, PDE1B, PSEN1, NF1, TH, PARK2
Regulation of neuron apoptotic process	7	9.19E−07	1.63E−03	PSEN1, HMOX1, BCL2, NF1, MECP2, PARK2, SIGMAR1
Response to decreased oxygen levels	8	9.52E−07	1.69E−03	HMOX1, BCL2, VEGFA, NF1, TH, MECP2, IL1B, TRH
Neuron apoptotic process	7	1.22E−06	2.16E−03	APP, PSEN1, HMOX1, BCL2, NF1, MECP2, PARK2
Response to oxygen levels	8	1.41E−06	2.51E−03	HMOX1, BCL2, VEGFA, NF1, TH, MECP2, IL1B, TRH
Central nervous system development	11	1.65E−06	2.93E−03	SLC17A7, APP, APOD, PSEN1, MAPT, BCL2, NF1, TH, MECP2, IL1B, PARK2
Nitrogen compound transport	10	2.10E−06	3.74E−03	SLC17A7, SLC17A6, PSEN1, NF1, IL1RN, TH, MECP2, IL1B, PARK2, TRH
Learning	6	3.41E−06	6.06E−03	APP, PDE1B, NF1, TH, MECP2, PARK2
Response to oxidative stress	8	4.34E−06	7.72E−03	APP, APOD, PSEN1, MAPT, HMOX1, BCL2, IL1B, PARK2
Regulation of neuron death	7	5.01E−06	8.91E−03	PSEN1, HMOX1, BCL2, NF1, MECP2, PARK2, SIGMAR1
Chemical homeostasis	11	5.20E−06	9.25E−03	SLC17A7, MICU1, APP, PSEN1, KL, HMOX1, BCL2, VEGFA, TH, IL1B, PARK2
CC				
Synapse part	13	4.60E−10	5.87E−07	ACHE, TH, NF1, MECP2, PARK2, SIGMAR1, SYP, SLC17A7, APP, SLC17A6, PSEN1, GRIA1, MAPT
Neuron projection	15	1.09E−09	1.39E−06	TH, NF1, PARK2, SIGMAR1, SYP, SLC17A7, APP, HTR1A, SLC17A6, PNOC, APOD, PSEN1, GRIA1, MAPT, HTR7
Neuron part	16	4.99E−09	6.37E−06	TH, NF1, PARK2, SIGMAR1, SYP, SLC17A7, APP, HTR1A, SLC17A6, PNOC, PDE1B, APOD, PSEN1, GRIA1, MAPT, HTR7
Synapse	13	6.19E−09	7.90E−06	ACHE, TH, NF1, MECP2, PARK2, SIGMAR1, SYP, SLC17A7, APP, SLC17A6, PSEN1, GRIA1, MAPT
Presynapse	9	5.66E−08	7.23E−05	SYP, SLC17A7, APP, SLC17A6, PSEN1, GRIA1, NF1, TH, PARK2
Dendrite	10	2.30E−07	2.94E−04	APP, HTR1A, PNOC, APOD, PSEN1, GRIA1, HTR7, MAPT, NF1, TH
Somatodendritic compartment	11	4.77E−07	6.09E−04	APP, HTR1A, PNOC, APOD, PDE1B, PSEN1, GRIA1, HTR7, MAPT, NF1, TH

Term	Count	P value	FDR	Genes
Axon	9	1.38E–06	1.76E–03	SYP, SLC17A7, APP, PNOC, PSEN1, GRIA1, MAPT, NF1, TH
Secretory vesicle	9	2.35E–06	3.00E–03	SYP, SLC17A7, APP, SLC17A6, GRIA1, VEGFA, TH, IL1B, TRH
MF				
Protein domain specific binding	10	7.72E–07	1.04E–03	SYP, APP, PSEN1, GRIA1, MAPT, BCL2, TH, MECP2, IL1B, PARK2

Table 3. GO analysis for the genes related to learning disorders.

Terms	Count	P value	Genes
KEGG			
Nicotine addiction	3	0.0047	SLC17A7, SLC17A6, GRIA1
Alzheimer's disease	4	0.0088	APP, PSEN1, MAPT, IL1B
Retrograde endocannabinoid signalling	3	0.0278	SLC17A7, SLC17A6, GRIA1
Serotonergic synapse	3	0.0331	APP, HTR1A, HTR7
Glutamatergic synapse	3	0.0348	SLC17A7, SLC17A6, GRIA1
WikiPathways			
SIDS susceptibility pathways	6	2.83E–05	MECP2, IL1RN, TH, IL1B, HTR1A, VEGFA
Integrated pancreatic cancer pathway	5	0.0008	APP, ACHE, BCL2, MAPT, VEGFA
Monoamine transport	3	0.0003	ACHE, TH, IL1B
Synaptic vesicle pathway	3	0.0011	SLC17A6, SLC17A7, SYP
Alzheimer's disease	4	0.0011	APP, IL1B, PSEN1, MAPT
Mecp2 and associated Rett syndrome	3	0.0009	GRIA1, MECP2, NF1
Overview of nanoparticle effects	2	0.0028	BCL2, HMOX1
Reactome			
Organic anion transporters	2	0.0215	SLC17A7, SLC17A6
Serotonin receptors	2	0.0321	HTR1A, HTR7
TRAF6-mediated NF-kB activation	2	0.0631	APP, RNF135
Interleukin-1 signalling	2	0.0932	IL1RN, IL1B
Biocart			
Generation of amyloid b-peptide by PS1	2	0.0427	APP, PSEN1
Deregulation of CDK5 in Alzheimer's Disease	2	0.0427	APP, MAPT

Table 4. Pathway analysis for the genes related to learning disorders.

Only one module was found for LD (**Figure 2**: LD). In this module, HTR1A, MAPT, TH, PARK2 and TRH were interacted by physical interactions, co-expression, predicted, pathway, co-localization, genetic interactions and shared protein domains. PARK2 is known to play a role in neurological development or function and when disturbed can account for LD [38, 39]. TH links PARK2 by sharing the same pathway [40], in which HOXA4 could co-express with TRHR [41], which could co-express with MAPT [42]. Serotonin gene (HTR1A) is also involved in this module, which is consistent with the serotonergic synapse and serotonin receptors pathway in KEGG and Reactome, respectively.

4.3. Schizophrenia

A total of 149 genes related to Sch were found to be BPA bi-interacted. Significant BPs included those found in LD such as behaviour, learning or memory, cognition, synaptic signalling related and cell-cell signalling related. Synaptic transmission-related, cell communication-related and phosphorus metabolic processes were also found (**Table 5**). Like LD, the genes participate in synapse, presynapse, dendrite, somatodendritic compartment and axon cellular components. Other CCs include cell body, intrinsic or integral component of plasma membrane, synaptic membrane and plasma membrane region. Unlike LD with little significant MFs found, Sch showed many significant MFs including the activity of signal transducer, neurotransmitter receptor, transmembrane-signalling receptor, molecular transducer, glutamate receptor and dopamine neurotransmitter receptor, dopamine or catecholamine binding. The same as LD, WikiPathways found that BPA could be linked to Sch through Alzheimer's disease, monoamine transport and SIDS susceptibility pathways (**Table 6**). The KEGG pathways such as neuroactive ligand-receptor interaction, cocaine addiction, dopaminergic synapse, cAMP signalling pathway and calcium-signalling pathway were also found to be significant. Reactome pathways included transmission across chemical synapse, amine ligand-binding receptors, neuronal system and signalling by GPCR, PDGF, FGFR4, FGFR3, FGFR1 or EGFR.

Five molecular modules were found for Sch (**Figure 3**). In module 1, 24 genes were connected by predicted, co-expression, physical interaction, co-localization, shared protein domains and pathway and genetic interactions. Almost all these genes showed co-expression [43, 44] and genetic interactions [45]. Except these 24 genes, GRID2, GRIK3 and GRIK4 might also be influenced by BPA because of their shared protein domains with GRIK2, GRIK5, GRIN2D, GRM2 and GRM3 and genetic interactions. KCNJ12 might also be interacted by BPA because of predicted, co-expression, genetic interactions and physical interactions. KCNJ12 has been reported may involve in the candidate pathway of Sch [46]. IL6, HP, AKT1, GSK3B, TNF, PIK3CB, and TP53 are composites of module 2. AKT/GSK3 pathway in which AKT1 and GSK3B has been reportedly associated in Sch [47]. TP53, as a key element in maintaining genomic stability and cell apoptosis and having been evidently proved a Sch susceptibility gene, linked TNF, AKT1 and GSK3B by direct co-expression, physical interactions and genetic interactions. Other genes such as AKT2 and DVL1 in this module might also interact with BPA because of the co-expression of AKT2 and PIK3CB [37], and DVL1 and AKT1 [48]. In module 3, SLC6A3, GAD1, COMT and RELN were involved through physical interaction, co-expression, predicted, pathway, co-localization and shared protein domains. LRPAP1, ITGB1, DAB1, PAFAH1B3,

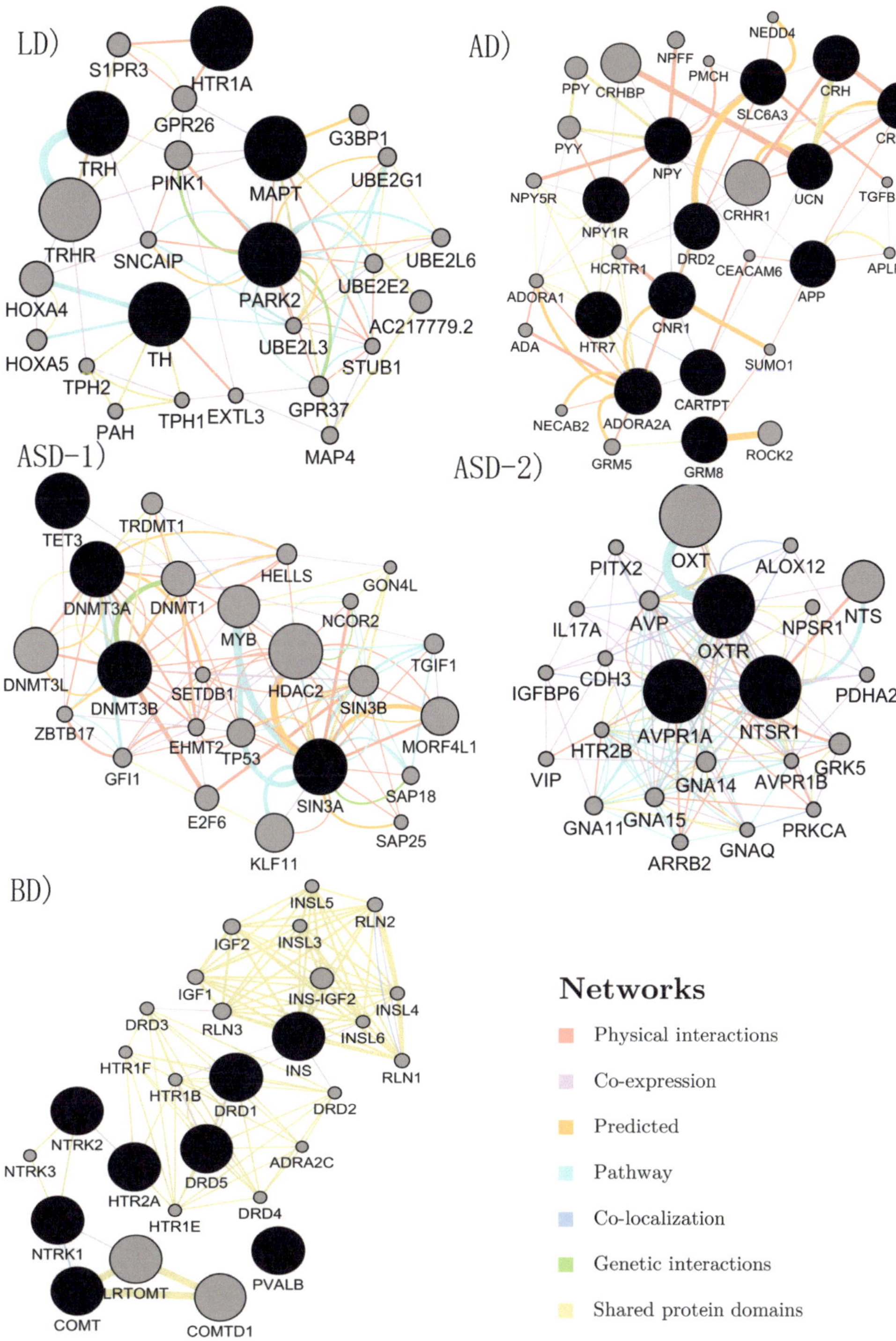

Figure 2. Networks for the genes in the PPI MCODE molecular modules for LD, ASD, AD and BD.

Term	Count	P value	FDR	Genes
BP				
Synaptic signalling	44	2.29E−27	4.33E−24	DRD1, CPLX2, CCL2, CPLX1, SLC6A1, DRD3, GABRB2, GRIK2, DRD2, SLC6A3, SLC6A4, DRD4, GRIK5, OXTR, TAC1, DTNBP1, AKT1, SYP, PLCL2, GAD2, HRH1, GRIN2B, APOE, GRIN2D, CNR1, SYN2, CAMK2B, GAD1, DLG1, GABRD, NOS1, NRXN1, NTSR1, LAMA2, GRM3, GRM2, GSK3A, HTR7, GSK3B, NTRK1, HTR6, RELN, NRGN, HTR2A
Anterograde Trans-synaptic signalling	44	2.29E−27	4.33E−24	DRD1, CPLX2, CCL2, CPLX1, SLC6A1, DRD3, GABRB2, GRIK2, DRD2, SLC6A3, SLC6A4, DRD4, GRIK5, OXTR, TAC1, DTNBP1, AKT1, SYP, PLCL2, GAD2, HRH1, GRIN2B, APOE, GRIN2D, CNR1, SYN2, CAMK2B, GAD1, DLG1, GABRD, NOS1, NRXN1, NTSR1, LAMA2, GRM3, GRM2, GSK3A, HTR7, GSK3B, NTRK1, HTR6, RELN, NRGN, HTR2A
Trans-synaptic signalling	44	2.29E−27	4.33E−24	DRD1, CPLX2, CCL2, CPLX1, SLC6A1, DRD3, GABRB2, GRIK2, DRD2, SLC6A3, SLC6A4, DRD4, GRIK5, OXTR, TAC1, DTNBP1, AKT1, SYP, PLCL2, GAD2, HRH1, GRIN2B, APOE, GRIN2D, CNR1, SYN2, CAMK2B, GAD1, DLG1, GABRD, NOS1, NRXN1, NTSR1, LAMA2, GRM3, GRM2, GSK3A, HTR7, GSK3B, NTRK1, HTR6, RELN, NRGN, HTR2A
Chemical synaptic transmission	44	2.29E−27	4.33E−24	DRD1, CPLX2, CCL2, CPLX1, SLC6A1, DRD3, GABRB2, GRIK2, DRD2, SLC6A3, SLC6A4, DRD4, GRIK5, OXTR, TAC1, DTNBP1, AKT1, SYP, PLCL2, GAD2, HRH1, GRIN2B, APOE, GRIN2D, CNR1, SYN2, CAMK2B, GAD1, DLG1, GABRD, NOS1, NRXN1, NTSR1, LAMA2, GRM3, GRM2, GSK3A, HTR7, GSK3B, NTRK1, HTR6, RELN, NRGN, HTR2A
Cell-cell signalling	62	3.34E−26	6.31E−23	SLC6A1, FAM3D, GRIK2, GABRB2, SLC6A3, SLC6A4, GRIK5, VIPR2, LGR4, AKT1, SYP, BDNF, GRIN2B, APOE, GRIN2D, IL1B, PLCB1, HCAR2, DISC1, DLG1, AVP, MAGI2, NRXN1, NTSR1, GRM3, GRM2, HTR7, HTR6, RELN, NRGN, FGFR1, DRD1, CPLX2, CCL2, CPLX1, TNF, DRD3, HLA-DRB1, DRD2, DRD4, OXTR, TAC1, DTNBP1, PLCL2, HRH1, GAD2, CNR1, SYN2, CAMK2B, VPS35, GAD1, GABRD, DIXDC1, IL6, NOS1, NTF3, LAMA2, LRP1, GSK3A, NTRK1, GSK3B, HTR2A
Modulation of synaptic transmission	31	6.10E−24	1.15E−20	CPLX2, DRD1, CCL2, DRD3, SLC6A1, GRIK2, DRD2, DRD4, SLC6A4, GRIK5, TAC1, OXTR, DTNBP1, SYP, PLCL2, HRH1, APOE, CNR1, CAMK2B, NOS1, NTF3, NRXN1, NTSR1, LAMA2, GRM3, GRM2, GSK3B, NTRK1, RELN, NRGN, HTR2A
Regulation of cell communication	72	2.46E−17	4.64E−14	SLC6A1, FAM3D, GRIK2, SLC6A4, GRIK5, LGR4, SYP, AKT1, BDNF, PAK2, APOE, IL1B, PLCB1, NRG1, HCAR2, DISC1, ALS2CL, DLG1, AVP, MAGI2, PIK3CB, TP53, NRXN1, IL6R, NTSR1, GRM3, GRM2, HTR6, RELN, NRGN, ADAMTS3, FGFR1, DRD1, CPLX2, CCL2, TNF, DRD3, HLA-DRB1, DRD2, DRD4, COL3A1, PML, TAC1, OXTR, NPRL2, DTNBP1, PLCL2, HRH1, RGS12, RB1CC1, CNR1, CAMK2B, VPS35, THBS1, CHD4, DIXDC1, IL6, NOS1, NTF3, PHB, RTN4R, CHI3L1, KDR, LAMA2, LRP1, GSK3A, NTRK1, RGS4, GSK3B, MTOR, RGS9, HTR2A

Term	Count	P value	FDR	Genes
Regulation of signalling	72	5.99E−17	2.11E−13	SLC6A1, FAM3D, GRIK2, SLC6A4, GRIK5, LGR4, SYP, AKT1, BDNF, PAK2, APOE, IL1B, PLCB1, NRG1, HCAR2, DISC1, ALS2CL, DLG1, AVP, MAGI2, PIK3CB, TP53, NRXN1, IL6R, NTSR1, GRM3, GRM2, HTR6, RELN, NRGN, ADAMTS3, FGFR1, DRD1, CPLX2, CCL2, TNF, DRD3, HLA-DRB1, DRD2, DRD4, COL3A1, PML, TAC1, OXTR, NPRL2, DTNBP1, PLCL2, HRH1, RGS12, RB1CC1, CNR1, CAMK2B, VPS35, THBS1, CHD4, DIXDC1, IL6, NOS1, NTF3, PHB, RTN4R, CHI3L1, KDR, LAMA2, LRP1, GSK3A, NTRK1, RGS4, GSK3B, MTOR, RGS9, HTR2A
Single-organism behaviour	27	1.71E−15	3.14E−12	DRD1, DRD3, SLC6A1, DRD2, GRIK2, DRD4, SLC6A4, TAC1, OXTR, COMT, HRH1, GRIN2B, GRIN2D, CNR1, IL1B, THBS1, PLCB1, AVP, IL6, NOS1, NRXN1, NTSR1, NTRK1, RELN, MTOR, NRGN, HTR2A
Positive regulation of cell communication	48	9.61E−15	1.82E−11	FGFR1, DRD1, CCL2, TNF, DRD3, HLA-DRB1, GRIK2, DRD2, DRD4, COL3A1, PML, OXTR, TAC1, DTNBP1, LGR4, AKT1, BDNF, PAK2, RB1CC1, IL1B, VPS35, NRG1, PLCB1, THBS1, HCAR2, DISC1, DIXDC1, IL6, NOS1, NTF3, PIK3CB, PHB, TP53, CHI3L1, IL6R, NRXN1, NTSR1, KDR, LAMA2, GSK3A, NTRK1, GSK3B, HTR6, RELN, NRGN, MTOR, ADAMTS3, HTR2A
Positive regulation of signalling	48	1.10E−14	2.08E−11	FGFR1, DRD1, CCL2, TNF, DRD3, HLA-DRB1, GRIK2, DRD2, DRD4, COL3A1, PML, OXTR, TAC1, DTNBP1, LGR4, AKT1, BDNF, PAK2, RB1CC1, IL1B, VPS35, NRG1, PLCB1, THBS1, HCAR2, DISC1, DIXDC1, IL6, NOS1, NTF3, PIK3CB, PHB, TP53, CHI3L1, IL6R, NRXN1, NTSR1, KDR, LAMA2, GSK3A, NTRK1, GSK3B, HTR6, RELN, NRGN, MTOR, ADAMTS3, HTR2A
Learning or memory	21	1.10E−14	2.08E−11	DRD1, DRD3, SLC6A1, DRD2, DRD4, SLC6A4, TAC1, OXTR, COMT, NRXN1, NTSR1, HRH1, GRIN2B, CNR1, NTRK1, IL1B, RELN, NRGN, MTOR, PLCB1, HTR2A
Behaviour	30	1.11E−14	2.10E−11	DRD1, DRD3, SLC6A1, DRD2, GRIK2, SLC6A3, DRD4, SLC6A4, TAC1, OXTR, COMT, HRH1, GRIN2B, GRIN2D, CNR1, IL1B, THBS1, PLCB1, NRG1, IL6, AVP, NOS1, TP53, NRXN1, NTSR1, NTRK1, RELN, MTOR, NRGN, HTR2A
Cognition	22	1.22E−14	2.31E−11	DRD1, DRD3, SLC6A1, DRD2, DRD4, SLC6A4, TAC1, OXTR, COMT, NRXN1, NTSR1, DGCR2, HRH1, GRIN2B, CNR1, NTRK1, IL1B, RELN, NRGN, MTOR, PLCB1, HTR2A
Neuron-neuron synaptic transmission	16	1.33E−14	2.52E−11	DRD1, DRD3, SLC6A1, DRD2, GRIK2, GABRB2, SLC6A3, DRD4, SLC6A4, GRIK5, TAC1, OXTR, NRXN1, NTRK1, RELN, HTR2A
Response to alkaloid	18	1.48E−14	2.79E−11	IL6, AVP, DRD1, TNF, NOS1, SLC6A1, ND4, DRD3, DRD2, SLC6A3, DRD4, TAC1, OXTR, CNR1, NTRK1, IL1B, MTOR, HTR2A
Positive regulation of synaptic transmission	16	8.02E−14	1.52E−10	DRD1, CCL2, NOS1, DRD2, GRIK2, DRD4, TAC1, OXTR, NRXN1, NTSR1, DTNBP1, LAMA2, NTRK1, GSK3B, RELN, NRGN
Regulation of phosphate metabolic process	47	2.16E−13	4.07E−10	FGFR1, DRD1, CCL2, NRG3, TNF, ADCY7, DRD3, GRIK2, DRD2, DRD4, PML, NPRL2, VIPR2, DTNBP1, PLCL2, AKT1, HRH1, PAK2, APOE, RB1CC1, IL1B, NRG1, PLCB1, THBS1, DLG1, IL6, AVP, MAGI2, NOS1, NTF3, PIK3CB, PHB, TP53, RTN4R, CHI3L1, IL6R, NTSR1, KDR, GRM3, GRM2, GSK3A, RGS4, NTRK1, GSK3B, RELN, MTOR, HTR2A

Term	Count	P value	FDR	Genes
Regulation of phosphorus metabolic process	47	2.20E−13	4.16E−10	FGFR1, DRD1, CCL2, NRG3, TNF, ADCY7, DRD3, GRIK2, DRD2, DRD4, PML, NPRL2, VIPR2, DTNBP1, PLCL2, AKT1, HRH1, PAK2, APOE, RB1CC1, IL1B, NRG1, PLCB1, THBS1, DLG1, IL6, AVP, MAGI2, NOS1, NTF3, PIK3CB, PHB, TP53, RTN4R, CHI3L1, IL6R, NTSR1, KDR, GRM3, GRM2, GSK3A, RGS4, NTRK1, GSK3B, RELN, MTOR, HTR2A
Regulation of transport	49	8.08E−13	1.53E−09	FGFR1, DRD1, CPLX2, CCL2, CPLX1, TNF, SLC6A1, DRD3, HLA-DRB1, FAM3D, DRD2, DRD4, GRIK5, PML, OXTR, TAC1, COMT, EDEM2, DTNBP1, LGR4, AKT1, APOE, CNR1, PDE4B, IL1B, CAMK2B, VPS35, NRG1, THBS1, HCAR2, DLG1, IL6, AVP, MAGI2, NOS1, NTF3, PIK3CB, MAOB, TP53, AHI1, NRXN1, NTSR1, PCM1, LRP1, GSK3A, GSK3B, RELN, MTOR, HTR2A
CC				
Neuron projection	48	3.23E−21	4.54E−18	DRD1, CPLX2, CCL2, CPLX1, SLC6A1, GRIK2, DRD2, SLC6A3, SLC6A4, DRD4, GRIK5, TAC1, COMT, DTNBP1, HNRNPA3, SYP, GAD2, MTHFR, RGS12, GRIN2B, PVALB, APOE, CNR1, PDE4B, CAMK2B, NRG1, DISC1, DLG1, AVP, NOS1, MAGI2, RTN4R, NRXN1, NTSR1, LAMA2, GRM3, GRM2, LRP1, HTR7, NTRK1, GSK3B, HTR6, RELN, NRGN, MTOR, TPH1, KPNA1, HTR2A
Synapse	40	6.12E−19	8.61E−16	CPLX2, DRD1, CCL2, CPLX1, GABRB2, GRIK2, DRD2, SLC6A3, DRD4, SLC6A4, GRIK5, COMT, DTNBP1, AKT1, SYP, GAD2, MTHFR, RGS12, GRIN2B, PVALB, GRIN2D, PDE4B, SYN2, CAMK2B, NRG1, GAD1, DISC1, DLG1, GABRD, NOS1, MAGI2, GABRA6, NRXN1, NTSR1, LAMA2, GRM3, GRM2, GSK3A, GSK3B, NRGN
Synapse part	36	1.32E−18	1.86E−15	CPLX2, DRD1, CPLX1, GABRB2, DRD2, GRIK2, SLC6A3, DRD4, SLC6A4, GRIK5, COMT, DTNBP1, SYP, AKT1, GAD2, GRIN2B, PVALB, GRIN2D, PDE4B, SYN2, CAMK2B, GAD1, DISC1, DLG1, GABRD, NOS1, MAGI2, GABRA6, NRXN1, NTSR1, LAMA2, GRM3, GRM2, GSK3A, GSK3B, NRGN
Neuron part	51	3.07E−18	4.33E−15	SLC6A1, GRIK2, SLC6A3, SLC6A4, GRIK5, SYP, GRIN2B, PVALB, APOE, PDE4B, NRG1, DISC1, DLG1, AVP, MAGI2, NRXN1, NTSR1, GRM3, GRM2, HTR7, HTR6, RELN, NRGN, TPH1, KPNA1, CPLX2, DRD1, CPLX1, CCL2, DRD2, DRD4, TAC1, COMT, DTNBP1, HNRNPA3, GAD2, MTHFR, RGS12, CNR1, SYN2, CAMK2B, GAD1, NOS1, RTN4R, LAMA2, LRP1, NTRK1, GSK3B, RGS9, MTOR, HTR2A
Dendrite	32	1.49E−17	2.10E−14	CPLX2, DRD1, CCL2, CPLX1, GRIK2, DRD2, DRD4, GRIK5, COMT, DTNBP1, RGS12, APOE, PDE4B, CAMK2B, NRG1, AVP, NOS1, MAGI2, NTSR1, LAMA2, GRM3, GRM2, LRP1, GSK3B, NTRK1, HTR7, HTR6, RELN, NRGN, MTOR, KPNA1, HTR2A
Somatodendritic compartment	37	2.66E−17	3.74E−14	DRD1, CPLX2, CCL2, CPLX1, GRIK2, DRD2, SLC6A3, DRD4, GRIK5, TAC1, COMT, DTNBP1, RGS12, PVALB, APOE, PDE4B, CAMK2B, NRG1, AVP, NOS1, MAGI2, RTN4R, NRXN1, NTSR1, LAMA2, GRM3, GRM2, LRP1, HTR7, GSK3B, NTRK1, HTR6, RELN, NRGN, MTOR, KPNA1, HTR2A

Term	Count	P value	FDR	Genes
Axon	29	7.56E−16	1.10E−12	CPLX2, DRD1, CCL2, CPLX1, SLC6A1, DRD2, GRIK2, SLC6A3, DRD4, GRIK5, TAC1, COMT, DTNBP1, SYP, GAD2, PVALB, CNR1, NRG1, DISC1, DLG1, RTN4R, NRXN1, NTSR1, GRM3, GRM2, GSK3B, NTRK1, NRGN, HTR2A
Postsynapse	25	4.28E−14	6.02E−11	GABRD, DRD1, NOS1, MAGI2, GABRB2, GRIK2, DRD2, GABRA6, DRD4, GRIK5, COMT, NTSR1, DTNBP1, AKT1, LAMA2, GRM3, GRIN2B, GSK3A, GRIN2D, GSK3B, PDE4B, CAMK2B, NRGN, DISC1, DLG1
Cell body	26	2.58E−11	3.63E−08	CPLX2, DRD1, CPLX1, CCL2, DRD2, GRIK2, SLC6A3, DRD4, GRIK5, TAC1, COMT, RGS12, PVALB, APOE, CAMK2B, NRG1, DISC1, RTN4R, NRXN1, NTSR1, LRP1, NTRK1, GSK3B, MTOR, NRGN, HTR2A
Presynapse	20	2.97E−11	4.19E−08	CPLX2, CPLX1, DRD2, GRIK2, SLC6A3, DRD4, SLC6A4, GRIK5, NRXN1, NTSR1, DTNBP1, SYP, GAD2, GRM3, GRM2, PVALB, PDE4B, SYN2, GAD1, DISC1
Intrinsic component of plasma membrane	45	1.08E−10	1.52E−07	FGFR1, DRD1, NRG3, TNF, SLC6A1, DRD3, HLA-DRB1, GABRB2, GRIK2, DRD2, PLXNA2, SLC6A3, SLC6A4, DRD4, GRIK5, OXTR, VIPR2, LGR4, HRH1, GRIN2B, GRIN2D, CNR1, NRG1, DLG1, GABRD, IL6, GABRA6, PHB, MET, NTNG1, RTN4R, IL6R, NRXN1, NTSR1, TSPAN18, KDR, GRM3, GRM2, LRP1, SLC26A7, HTR7, NTRK1, HTR6, CP, HTR2A
Integral component of plasma membrane	43	4.17E−10	5.87E−07	FGFR1, DRD1, TNF, NRG3, SLC6A1, HLA-DRB1, DRD3, GABRB2, GRIK2, DRD2, PLXNA2, SLC6A3, SLC6A4, DRD4, GRIK5, OXTR, VIPR2, LGR4, HRH1, GRIN2B, GRIN2D, CNR1, NRG1, DLG1, GABRD, IL6, GABRA6, PHB, MET, RTN4R, IL6R, NRXN1, NTSR1, TSPAN18, KDR, GRM3, GRM2, LRP1, SLC26A7, HTR7, NTRK1, HTR6, HTR2A
Neuronal cell body	23	4.45E−10	6.27E−07	DRD1, CPLX2, CCL2, CPLX1, GRIK2, DRD2, SLC6A3, DRD4, GRIK5, RTN4R, TAC1, NRXN1, NTSR1, LRP1, RGS12, PVALB, APOE, GSK3B, NTRK1, CAMK2B, NRGN, MTOR, HTR2A
Axon part	16	4.61E−09	6.49E−06	DRD1, CPLX2, CCL2, CPLX1, DRD2, GRIK2, DRD4, RTN4R, GRIK5, NRXN1, NTSR1, DTNBP1, SYP, PVALB, NRG1, DLG1
Dendritic spine	12	2.08E−08	2.93E−05	LAMA2, DRD1, GRM3, NOS1, DRD2, GSK3B, PDE4B, DRD4, COMT, NRGN, NTSR1, DTNBP1
Neuron spine	12	2.46E−08	3.47E−05	LAMA2, DRD1, GRM3, NOS1, DRD2, GSK3B, PDE4B, DRD4, COMT, NRGN, NTSR1, DTNBP1
Synaptic membrane	16	4.23E−08	5.96E−05	GABRD, GRIK2, GABRB2, GABRA6, GRIK5, NRXN1, COMT, DTNBP1, SYP, GRM3, GAD2, GRM2, GRIN2B, GRIN2D, DISC1, DLG1
Excitatory synapse	13	8.32E−08	1.17E−04	SYP, GRM3, MAGI2, NOS1, DRD2, GRIK2, PDE4B, GRIK5, CAMK2B, NRGN, DTNBP1, DISC1, DLG1
Plasma membrane region	28	1.94E−07	2.73E−04	DRD1, DRD2, GRIK2, GABRB2, SLC6A3, GRIK5, OXTR, COMT, DTNBP1, SYP, GAD2, GRIN2B, GRIN2D, NRG1, DISC1, DLG1, GABRD, NOS1, GABRA6, MET, GIF, NRXN1, IL6R, GRM3, GRM2, SLC26A7, RGS9, HTR2A

Term	Count	P value	FDR	Genes
Axon terminus	11	2.90E−07	4.08E−04	SYP, DRD1, CPLX2, CPLX1, CCL2, PVALB, DRD2, GRIK2, DRD4, GRIK5, NTSR1
MF				
Signal transducer activity	46	2.55E−11	3.79E−08	FGFR1, DRD1, NRG3, DRD3, HLA-DRB1, GABRB2, GRIK2, DRD2, PLXNA2, ADGRF4, DRD4, GRIK5, OXTR, NR3C1, VIPR2, LGR4, PLCL2, TAAR6, HRH1, RGS12, GRIN2B, PAK2, GRIN2D, CNR1, PLCB1, NRG1, HCAR2, GABRD, AVP, MAGI2, IL2RA, GABRA6, MET, RTN4R, IL6R, NRXN1, NTSR1, KDR, GPR153, GRM3, GRM2, HTR7, NTRK1, HTR6, RGS9, HTR2A
Neurotransmitter receptor activity	12	8.07E−10	1.20E−06	DRD1, GRIN2B, DRD3, DRD2, GRIK2, HTR7, GRIN2D, GABRA6, HTR6, DRD4, GRIK5, HTR2A
Transmembrane receptor activity	37	2.01E−09	2.98E−06	FGFR1, DRD1, HLA-DRB1, DRD3, GABRB2, GRIK2, PLXNA2, DRD2, ADGRF4, DRD4, GRIK5, OXTR, VIPR2, LGR4, TAAR6, HRH1, GRIN2B, GRIN2D, CNR1, HCAR2, DLG1, GABRD, IL2RA, GABRA6, MET, RTN4R, IL6R, NRXN1, NTSR1, KDR, GPR153, GRM3, GRM2, NTRK1, HTR7, HTR6, HTR2A
Transmembrane signalling receptor activity	36	2.61E−09	3.87E−06	FGFR1, DRD1, HLA-DRB1, DRD3, GABRB2, GRIK2, PLXNA2, DRD2, ADGRF4, DRD4, GRIK5, OXTR, VIPR2, LGR4, HRH1, TAAR6, GRIN2B, GRIN2D, CNR1, HCAR2, GABRD, IL2RA, GABRA6, MET, RTN4R, IL6R, NRXN1, NTSR1, KDR, GPR153, GRM3, GRM2, NTRK1, HTR7, HTR6, HTR2A
Signalling receptor activity	37	5.46E−09	8.09E−06	FGFR1, DRD1, HLA-DRB1, DRD3, GABRB2, GRIK2, PLXNA2, DRD2, ADGRF4, DRD4, GRIK5, OXTR, NR3C1, VIPR2, LGR4, HRH1, TAAR6, GRIN2B, GRIN2D, CNR1, HCAR2, GABRD, IL2RA, GABRA6, MET, RTN4R, IL6R, NRXN1, NTSR1, KDR, GPR153, GRM3, GRM2, NTRK1, HTR7, HTR6, HTR2A
Molecular transducer activity	39	4.31E−08	6.38E−05	FGFR1, DRD1, HLA-DRB1, DRD3, GABRB2, GRIK2, PLXNA2, DRD2, ADGRF4, DRD4, GRIK5, OXTR, NR3C1, VIPR2, LGR4, TAAR6, HRH1, GRIN2B, GRIN2D, CNR1, HCAR2, DLG1, GABRD, IL2RA, GABRA6, MET, RTN4R, IL6R, NRXN1, NTSR1, KDR, GPR153, GRM3, LRP1, GRM2, NTRK1, HTR7, HTR6, HTR2A
Receptor activity	39	4.31E−08	6.38E−05	FGFR1, DRD1, HLA-DRB1, DRD3, GABRB2, GRIK2, PLXNA2, DRD2, ADGRF4, DRD4, GRIK5, OXTR, NR3C1, VIPR2, LGR4, TAAR6, HRH1, GRIN2B, GRIN2D, CNR1, HCAR2, DLG1, GABRD, IL2RA, GABRA6, MET, RTN4R, IL6R, NRXN1, NTSR1, KDR, GPR153, GRM3, LRP1, GRM2, NTRK1, HTR7, HTR6, HTR2A
Receptor binding	34	4.69E−07	6.95E−04	CCL2, NRG3, TNF, FAM3D, DRD3, DRD2, SLC6A3, COL3A1, TAC1, PLCL2, BDNF, APOE, TRAK1, IL1B, VPS35, DAO, NRG1, THBS1, DLG1, IL6, AVP, MAGI2, NTF3, PHB, TP53, IL6R, NRXN1, KDR, NRIP1, LAMA2, LAMA1, LRP1, NTRK1, RELN
Dopamine binding	5	1.93E−06	2.85E−03	DRD1, DRD3, DRD2, SLC6A3, DRD4
Glutamate receptor activity	6	3.67E−06	5.44E−03	GRM3, GRM2, GRIN2B, GRIK2, GRIN2D, GRIK5

Term	Count	P value	FDR	Genes
Drug binding	9	8.84E−06	1.31E−02	DRD1, IL2RA, DRD3, DRD2, SLC6A3, CNR1, SLC6A4, DRD4, HTR2A
Catecholamine binding	5	1.03E−05	1.52E−02	DRD1, DRD3, DRD2, SLC6A3, DRD4
Dopamine neurotransmitter receptor activity	4	2.39E−05	3.54E−02	DRD1, DRD3, DRD2, DRD4

Table 5. GO analysis for the genes related to schizophrenia.

Term	Count	*P* value	Genes
KEGG			
Neuroactive ligand-receptor interaction	23	2.33E−08	GABRD, DRD1, DRD3, GABRB2, GRIK2, DRD2, GABRA6, DRD4, GRIK5, OXTR, NR3C1, NTSR1, VIPR2, HRH1, GRM3, TAAR6, GRM2, GRIN2B, GRIN2D, CNR1, HTR7, HTR6, HTR2A
Cocaine addiction	10	2.35E−05	DRD1, GRM3, BDNF, GRM2, GRIN2B, DRD2, SLC6A3, GRIN2D, MAOB, RGS9
Dopaminergic synapse	13	2.28E−04	DRD1, DRD3, DRD2, SLC6A3, MAOB, DRD4, COMT, AKT1, GRIN2B, GSK3A, GSK3B, CAMK2B, PLCB1
cAMP signalling pathway	14	4.17E−03	DRD1, ADCY7, PIK3CB, DRD2, OXTR, VIPR2, AKT1, BDNF, GRIN2B, GRIN2D, PDE4B, HTR6, CAMK2B, HCAR2
Calcium signalling pathway	12	4.66E−02	DRD1, HRH1, NOS1, ADCY7, HTR7, GRIN2D, HTR6, OXTR, CAMK2B, NTSR1, PLCB1, HTR2A
WikiPathways			
Monoamine GPCRs	8	3.30E−05	HTR6, HRH1, HTR7, HTR2A, DRD1, DRD2, DRD3, DRD4
Circadian rhythm-related genes	15	9.90E−05	NTRK1, GSK3B, TPH1, PML, SLC6A4, IL6, HTR7, NRIP1, AVP, DRD1, DRD2, TP53, DRD3, LGR4, DRD4
SIDS susceptibility pathways	12	8.30E−04	IL6, TPH1, VIPR2, BDNF, IL1B, HTR2A, AVP, TAC1, NR3C1, TNF, IL6R, SLC6A4
Spinal cord injury	10	9.54E−04	RTN4R, IL6, BDNF, IL1B, PLXNA2, CCL2, NOS1, TP53, TNF, RTN4
Alzheimer's disease	10	9.54E−04	GSK3B, LRP1, IL1B, APOE, NOS1, PLCB1, TP53, TNF, GRIN2B, GRIN2D
Vitamin B12 metabolism	7	9.54E−04	IL6, GIF, IL1B, MTHFR, CCL2, APOE, TNF
Monoamine transport	6	8.30E−04	IL1B, NOS1, SLC6A1, TNF, SLC6A3, SLC6A4
Hypothetical network for drug addiction	6	8.30E−04	CAMK2B, DRD1, DRD2, GRIN2B, DRD4, GRIN2D
Reactome			
Transmission across chemical synapses	17	7.71E−07	CAMK2B, GABRB2, GABRA6, GRIK5, GAD1, GAD2, GRIK2, SLC6A1, COMT, ADCY7, GRIN2B, SYN2, CPLX1, SLC6A3, GRIN2D, DLG1, PLCB1

Term	Count	*P* value	Genes
Amine ligand-binding receptors	9	2.36E−06	HTR6, HRH1, HTR7, TAAR6, HTR2A, DRD1, DRD2, DRD3, DRD4
Neuronal system	17	3.63E−05	CAMK2B, GABRB2, GABRA6, GRIK5, GAD1, GAD2, GRIK2, SLC6A1, COMT, ADCY7, GRIN2B, SYN2, CPLX1, SLC6A3, GRIN2D, DLG1, PLCB1
Signalling by GPCR	35	2.07E−04	CAMK2B, OXTR, VIPR2, PIK3CB, HTR2A, PHB, ADCY7, HCAR2, RGS4, GRM3, GRM2, HTR6, HRH1, HTR7, CNR1, PDE4B, AKT1, CCL2, DRD1, TAC1, DRD2, RGS9, DRD3, DRD4, NTSR1, TAAR6, NRG1, GRIN2B, GRIN2D, NRG3, IL2RA, RGS12, AVP, PLCB1, FGFR1
Signal Transduction	53	2.07E−04	GSK3B, GSK3A, OXTR, VIPR2, TRRAP, HTR2A, PIK3CB, NR3C1, TNF, GRM3, RGS4, GRM2, HTR6, HTR7, KDR, PDE4B, AKT1, VPS35, IL6R, RGS9, TAAR6, NRG1, NRG3, AVP, PLCB1, TP53, CAMK2B, LRP1, PHB, ADCY7, THBS1, RTN4, HCAR2, HRH1, CNR1, CCL2, APOE, DRD1, DRD2, TAC1, DRD3, PAK2, DRD4, NTSR1, NTRK1, GRIN2B, MTOR, GRIN2D, IL6, IL2RA, RGS12, LGR4, FGFR1
Class A/1 (Rhodopsin-like receptors)	16	2.07E−04	OXTR, TAAR6, HTR2A, HCAR2, HTR6, HRH1, HTR7, CNR1, CCL2, AVP, DRD1, DRD2, TAC1, DRD3, NTSR1, DRD4
GPCR ligand binding	18	6.14E−04	OXTR, TAAR6, HTR2A, HCAR2, GRM3, GRM2, HTR6, HRH1, HTR7, CNR1, CCL2, DRD1, AVP, DRD2, TAC1, DRD3, DRD4, NTSR1
Signalling by PDGF	15	1.72E−03	CAMK2B, GSK3B, GSK3A, NRG1, PHB, PIK3CB, ADCY7, GRIN2B, THBS1, MTOR, GRIN2D, NRG3, IL2RA, AKT1, FGFR1
Downstream signalling of activated FGFR4	14	1.72E−03	CAMK2B, GSK3B, GSK3A, NRG1, PHB, PIK3CB, ADCY7, GRIN2B, MTOR, GRIN2D, NRG3, IL2RA, AKT1, FGFR1
Downstream signalling of activated FGFR3	14	1.72E−03	CAMK2B, GSK3B, GSK3A, NRG1, PHB, PIK3CB, ADCY7, GRIN2B, MTOR, GRIN2D, NRG3, IL2RA, AKT1, FGFR1
Signalling by FGFR4	14	1.72E−03	CAMK2B, GSK3B, GSK3A, NRG1, PHB, PIK3CB, ADCY7, GRIN2B, MTOR, GRIN2D, NRG3, IL2RA, AKT1, FGFR1
Downstream signalling of activated FGFR2	14	1.72E−03	CAMK2B, GSK3B, GSK3A, NRG1, PHB, PIK3CB, ADCY7, GRIN2B, MTOR, GRIN2D, NRG3, IL2RA, AKT1, FGFR1
Signalling by FGFR3	14	1.72E−03	CAMK2B, GSK3B, GSK3A, NRG1, PHB, PIK3CB, ADCY7, GRIN2B, MTOR, GRIN2D, NRG3, IL2RA, AKT1, FGFR1
Downstream signalling of activated FGFR1	14	1.72E−03	CAMK2B, GSK3B, GSK3A, NRG1, PHB, PIK3CB, ADCY7, GRIN2B, MTOR, GRIN2D, NRG3, IL2RA, AKT1, FGFR1
NGF signalling via TRKA from the plasma membrane	15	1.72E−03	NTRK1, CAMK2B, GSK3B, GSK3A, NRG1, PHB, PIK3CB, ADCY7, GRIN2B, MTOR, GRIN2D, NRG3, IL2RA, AKT1, FGFR1
Signalling by FGFR1	14	1.80E−03	CAMK2B, GSK3B, GSK3A, NRG1, PHB, PIK3CB, ADCY7, GRIN2B, MTOR, GRIN2D, NRG3, IL2RA, AKT1, FGFR1
DAP12 signalling	14	2.09E−03	CAMK2B, GSK3B, GSK3A, NRG1, PHB, PIK3CB, ADCY7, GRIN2B, MTOR, GRIN2D, NRG3, IL2RA, AKT1, FGFR1
Downstream signal transduction	14	2.00E−03	CAMK2B, GSK3B, GSK3A, NRG1, PHB, PIK3CB, ADCY7, GRIN2B, MTOR, GRIN2D, NRG3, IL2RA, AKT1, FGFR1

Term	Count	P value	Genes
Neurotransmitter receptor ainding and downstream transmission in the postsynaptic cell	10	7.83E−04	CAMK2B, GABRB2, DLG1, GABRA6, GRIK5, GRIK2, PLCB1, ADCY7, GRIN2B, GRIN2D
Signalling by EGFR	14	2.51E−03	CAMK2B, GSK3B, GSK3A, NRG1, PHB, PIK3CB, ADCY7, GRIN2B, MTOR, GRIN2D, NRG3, IL2RA,AKT1,FGFR1

Table 6. Pathway analysis for the genes related to schizophrenia.

PAFAH1B2, VLDLR, MAP1B and SNCA might interact with BPA because of their involvement in the same pathway with RELN [40]. LRTOMT and COMTD1 might also be the candidate interacted genes with BPA because of their shared protein domains. In module 4, 10 genes are involved. Co-expression and physical interactions are the main interaction modes for these genes. HCRTR1, ZDHHC23, CLIC6 and MUC20 are the possible candidate genes influenced by BPA mainly because of their physical and genetic interactions. In module 5, TSPAN18, NKAPL and ZKSCAN4 were connected by physical interactions, co-expression, co-localization and shared protein domains. NKAP, ZSCAN9, ZSCAN16-related genes could also be the candidate interacted genes of BPA because of the shared protein domains or co-expression.

4.4. Autism spectrum disorders

We found 51 ASD genes that could biinteract with BPA. These genes are partly involved in the BPs such as organism development, system development, synapse organization, behaviour, learning or memory, regulation of synapse structure or activity, multicellular organismal process, nervous system development, membrane potential, cellular process and cell-cell signalling-related processes. Involved CC was synaptic membrane, and MFs included neuroligin family protein binding and chromatin binding. The KEGG pathways such as neuroactive ligand-receptor interaction and cocaine addiction were found as Sch. Like ID and LD, MECP2 and associated Rett syndrome (WikiPathways) was also found in ASD. PRC2 methylates histones and DNA, non-integrin membrane-ECM interactions, and gastrin-CREB-signalling pathway via PKC and MAPK were the main pathways found in Reactome.

Two molecular modules were found for ASD (**Figure 2**: ASD-1, ASD-2). In module 1, TET3, SIN3A, DNMT3A and DNMT3B were connected to each other by physical interactions, co-expression, predicted, pathway, co-localization, genetic interactions and shared protein domains. HDAC2 and MORF4L1 might be the candidate genes interacted with BPA because of their predicted interaction with SIN3A [24]. DNMT3L and MYB are another two main genes in this module because of their direct or indirect interactions with other genes. AVPR1A, OXTR and NTSR1 composite the module 2 through physical interactions, co-expression, pathway, co-localization and shared protein domains. OXT and NTS are another two main genes in this module; they share the same pathway with OXTR, NTSR1, AVPR1A and some other genes [40] and, thus, might be influenced by BPA.

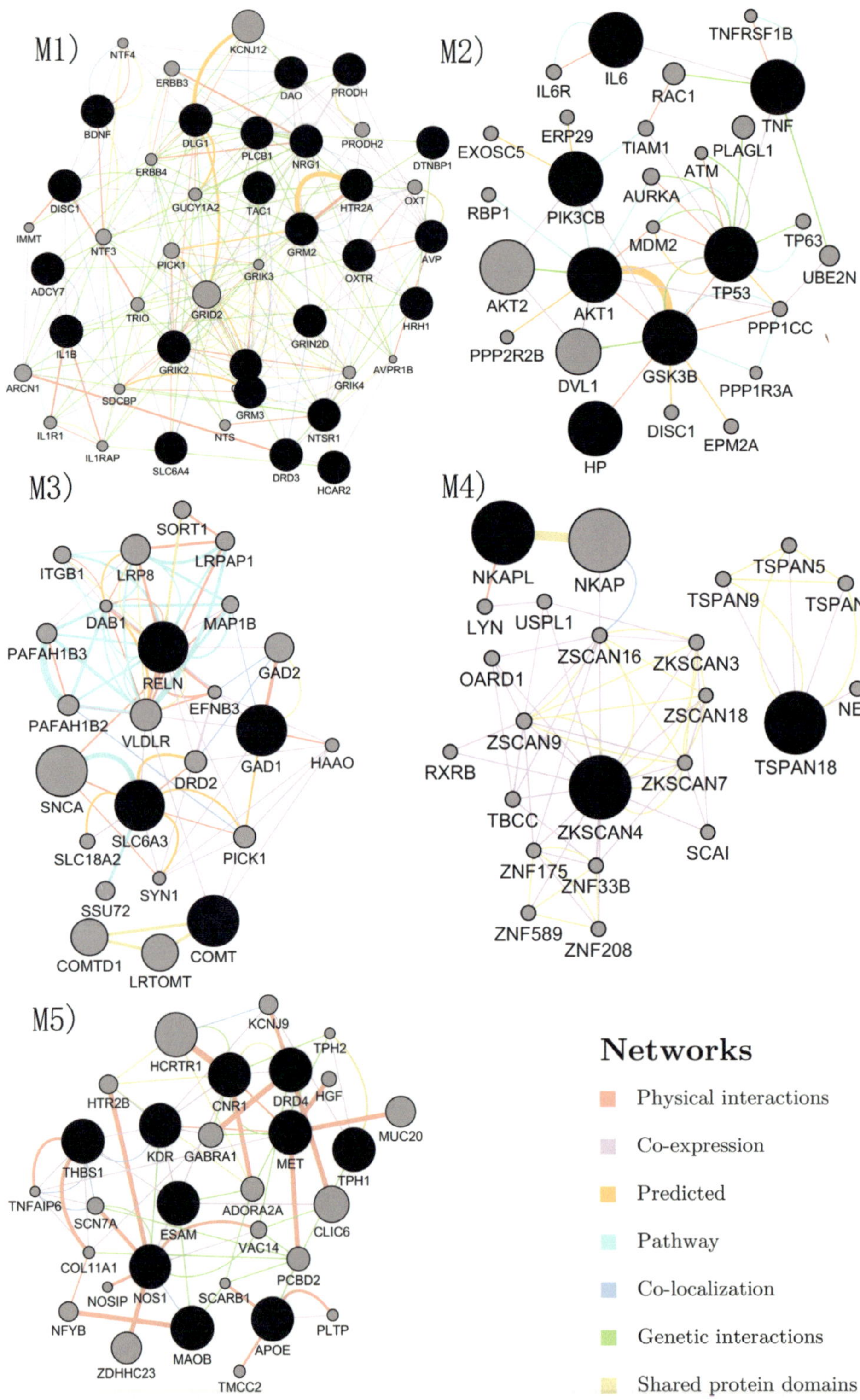

Figure 3. Networks for the genes in the PPI MCODE molecular modules for Sch.

4.5. Anxiety disorders

A total of 34 genes associated with AD were found bi-interacted with BPA. GO analysis for these genes indicated that behaviour, learning or memory, cognition, monoamine transport, chemical synaptic transmission, cell-cell signalling, anterograde trans-synaptic signalling and neurological system process were all significant BPs such as LD, Sch and ASD. Interestingly, blood circulation, circulatory system process and regulation of blood pressure were found significant for AD. Like ID, LD, Sch and ASD, significant CCs included neuron part, neuron projection, somatodendritic compartment, synapse part, axon, synapse, dendrite, cell body, presynapse and neuronal cell body. Significant MFs included receptor binding, neuropeptide hormone activity and G-protein–coupled receptor binding. The significant KEGG pathways were as those found in LD. Neuroactive ligand-receptor interaction, alcoholism, cAMP signalling pathway, serotonergic and dopaminergic synapse, Rap1 signalling pathway and retrograde endocannabinoid signalling are the potential KEGG pathways that might be influenced by BPA in AD. For WikiPathways, monoamine transport was again found related to BPA bi-interacted genes in AD. Other significant pathways included circadian rhythm-related genes, nicotine activity on dopaminergic neurons, corticotropin-releasing, GPCRs, cytosine methylation, myometrial relaxation and contraction pathways and estrogen-signalling pathway. In Reactome pathways, GPCR related, Class A/1, Class B/2, G alpha–related signalling events and peptide ligand-binding receptors were found possibly involved in the BPA-AD interactions.

Only one molecular module was found for AD, in which UCN, ADORA2A, CRH, NPS, CRHR2, NPY, NPY1R, APP, HTR7, CNR1, GRM8, SLC6A3, DRD2 and CARTPT were involved (**Figure 2**: AD). The interaction modes for these genes included predicted, physical interactions, shared protein domains, co-expression and co-localization. The genes in this module are all involved in neuroactive ligand-receptor interaction (KEGG pathway) which is consistent with the Reactome pathways of GPCR signalling and G alpha signalling events.

4.6. Bipolar disorder

A total of 39 genes were found bi-interacted with BPA for BD. The BPs, CCs and MFs were quite the same as AD. Neuroactive ligand-receptor interaction, dopaminergic synapse, calcium-signalling pathway, neurotrophin-signalling pathway, synaptic vesicle cycle, insulin secretion, morphine signalling pathway, MAPK signalling pathway, glutamatergic synapse and serotonergic synapse were found in KEGG pathways. Like LD and Sch, SIDS susceptibility pathways was also found significant for BD. GPCR-related pathways such as monoamine GPCRs in WikiPathways and GPCR ligand binding in Reactome were found to be involved in BPA-BD as in BPA-AD.

One molecular module was found for BD (**Figure 2**: BD), in which D1, NTRK1, DRD5, PVALB, NTRK2, HTR2A, COMT and INS were involved. Shared protein domains, co-localization and co-expression were the main interactions in this module. COMTD1 and LRTOMT might also be influenced by BPA because of their shared protein domains with COMT.

4.7. Other neurodevelopmental disorders

A total of 14 genes were found for bi-interacted BPA in DS. GO analysis indicated cellular oxidant detoxification-related BPs significant for these genes, and the MF of antioxidant activity was found significant consistently. The pathway analysis showed that KEGG pathway like one carbon pool by folate, and some pathways related to folate, one carbon or water-soluble vitamins metabolism in WikiPathways or Reactome pathways. Detoxification of reactive oxygen species and cellular responses to stress were also found significant in Reactome pathways. Consistent with the results of GO and pathway analyses, PPI interaction showed two different molecular modules, one with SLC19A1, MTR and MTHFR, and the other with SOD1, PRDX2 and PRDX6. It is clear that the module 1 is related to the clustering function of folate and other water-soluble vitamins metabolism, and the module 2 is for the detoxification of reactive oxygen species. Folate pathway has been regarded as involved in the pathogenesis of DS. Simultaneously, BPA exposure has the potential effects on the human phenotypes and altering DNA methylation [49, 50], which could be counteracted by the supplementation of methyl donors such as folate, choline, betaine and vitamin B12 [50]. Detoxification of reactive oxygen species and cellular responses to stress are important to maintain the mitochondrial function, which has been associated with the aetiology of early-onset dementia in patients with DS [51, 52].

For other NDs, less reference count or low inference score was found. But the limited results of GO and pathway analyses showed similar BPs, CCs, MFs and pathways with the above mentioned NDs in some extent.

5. Gene regulation

Transcription factors (TFs) and microRNAs (miRNAs), the largest families of transacting, share a common regulatory logic and represent the most numerous gene regulatory factors in multicellular genomes [53, 54]. The library of ENCODE and ChEA Consensus TFs from ChIP-X in EnrichR (http://amp.pharm.mssm.edu/Enrichr/ [13, 14]) were used for the possible TFs and related networks. The TargetScan library in EnrichR was used for the possible miRNA interaction. Here we only analysed the genes of ID, LD, Sch, ASD, LD, SMD, BD and SD whose inference score all over 10.

For the TFs, it was only ID, ASD, AD and BD that were found significant TFs (**Table 7**). USF2, MAX, SPI1, SMAD4, POU5F1, PPARD, MYC and RUNX1 were found significant for ID. The regulated genes for each of these TFs are shown in **Table 7**. The direct evidences for the USF2 linked to ID were the regulating role of USF2 on FMR1 of Fragile X mental retardation [55, 56]. SUZ12 was found common in ASD, AD and BD, and REST was found in both ASD and BD. SUZ12, as a component of the polycomb repressive complex, was shown to interact with some of long non-coding RNAs like AK055040 to involve in neural development and brain function [57]. REST is a key TF that represses expression of genes involved in neurogenesis and neuronal function in non-neural and immature neural cell types [58].

Some miRNAs were found in Sch, ASD, AD and BD (**Table 8**). MIR-218 and MIR-485-3p were significant in both Sch and AD. It has been reported that miR-218 is involved in Sch [59], and miR-485-3p is associated with obsessive-compulsive disorder, a type of AD [60]. MIR-380-3p was found significant in both ASD and BD, but no direct evidence in human studies.

Term	Overlap	Adjusted *P*-value	Combined score	Genes
ID				
USF2	16/965	0.0050	9.01	PECR, FMR1, PMM2, HEXA, PTEN, TRMT1, WDR62, ZBTB40, MED13L, AP4M1, NAGLU, CAPN10, FASN, SC5D, MAN1B1, RALGDS
MAX	27/2073	0.0049	8.74	HDAC4, KDM5A, HEXA, PTEN, ADK, WDR62, TSEN2, ZBTB40, AP4M1, EEF1B2, NAGLU, CASP2, SC5D, RALGDS, PARP1, PMM2, SRD5A3, ELP2, VRK1, TRMT1, TTI2, METTL23, AHI1, POLR3B, FASN, L2HGDH, MAN1B1
SPI1	17/1056	0.0050	8.46	KDM5A, KDM6B, TSEN34, DOCK8, DNMT3A, AP4E1, ELP2, VRK1, ERLIN2, TSEN54, TMCO1, EEF1B2, METTL23, AHI1, POLR3B, CAPN10, PEX6
SMAD4	11/584	0.0152	6.85	INPP4A, PDHX, SETBP1, DOCK8, PAX6, SRGAP3, FRY, MCC, GRIN2B, FGFR2, SHANK2
POU5F1	7/261	0.0167	6.31	EEF1B2, ENTPD1, UBR7, FASN, PAX6, FGFR2, ZBTB40
PPARD	7/285	0.0232	5.83	TMEM135, POLR3B, SLC31A1, PMM2, NF1, TSEN2, TTI2
MYC	18/1515	0.0416	4.47	ACBD6, PARP1, PMM2, SRD5A3, PTEN, ADK, SLC2A1, ELP2, TRMT1, TSEN2, AP4M1, EEF1B2, METTL23, POLR3B, NAGLU, FASN, MAN1B1, RALGDS
RUNX1	16/1294	0.0426	4.46	DOCK8, PTEN, SLC2A1, ELP2, FRY, KIF7, TSEN54, TMCO1, LETM1, NAGLU, DEAF1, CAPN10, NSD1, GNAS, SC5D, RALGDS
ASD				
SUZ12	15/1684	0.0009	10.75	DLX1, RYR2, OXTR, TSHZ3, BDNF, EN2, DIO3, NRXN2, AVPR1A, GRIN2B, DPP6, RELN, LRRTM3, SOX9, NTSR1
TCF3	9/1006	0.0323	5.60	LRRTM3, ITGB3, DNMT3A, DNMT3B, NRXN2, TBL1X, JARID2, FOXP1, SCN1A
REST	10/1280	0.0323	5.38	GABRB3, RYR2, DPP6, RELN, LRRN3, BDNF, NRXN1, DNMT3A, NRXN2, C3ORF58
AD				
SUZ12	13/1684	0.0001	14.64	GABRA2, EOMES, UCN, APP, CHRNA5, OXT, SLC6A4, HTR7, ADORA2A, CNR1, NPY, GRM8, DRD2
BD				
REST	10/1280	0.0073	8.07	POMC, SNAP25, NTRK2, RELN, TRPC3, BDNF, GRIK2, DRD1, CPLX2, DRD5
SUZ12	10/1684	0.0331	5.13	NTNG1, NTRK2, RELN, TENM4, BDNF, GRIK2, TAC1, CPLX2, DRD5, SLC6A4

Table 7. Transcription factors for the BPA-interacted genes involved in the neurodevelopmental disorders.

Term	Overlap	Adjusted P-value	Combined score	Genes
Sch				
MIR-485-3P	7/155	0.0171	7.68	ADAMTS3, CNR1, NRXN1, MAGI2, GAD2, CPLX2, NR3C1
MIR-218	11/402	0.0171	7.60	KLF12, RELN, HTR7, NRXN1, MAGI2, GRIK2, TAC1, NR3C1, SLC6A1, RTN4, LGR4
ASD				
MIR-380-3P	5/103	0.0008	13.50	MEF2C, LRRTM3, BDNF, NRXN1, IL1RAPL1
MIR-524	8/437	0.0009	12.82	DLX1, MEF2C, LRRTM3, PCDH9, IL1RAPL1, TBL1X, SOX9, FOXP1
MIR-368	3/40	0.0061	9.16	DLX1, MEF2C, BDNF
MIR-302C	5/243	0.0120	7.57	GABRB3, DLX1, PCDH9, TBL1X, FOXP1
MIR-518C	4/149	0.0143	7.07	PCDH9, ITGB3, DNMT3A, NRXN2
MIR-373	4/227	0.0483	5.01	GABRB3, DLG4, EN2, SOX9
MIR-191	2/29	0.0483	4.45	BDNF, FOXP1
AD				
MIR-218	5/402	0.0232	7.15	MECP2, HTR7, MAGI2, GNB1, NPY1R
MIR-498	3/114	0.0232	6.98	MECP2, CRH, PAM
MIR-101	4/257	0.0232	6.83	APP, MAGI2, GNB1, FOS
MIR-141, MIR-200A	4/310	0.0325	6.07	MECP2, CNR1, DIXDC1, DRD2
MIR-485-3P	3/155	0.0336	5.97	CNR1, MAGI2, GNB1
BD				
MIR-494	4/164	0.0251	6.71	SP4, GRIK2, CACNA1C, TAC1
MIR-410	3/93	0.0317	6.41	NTRK2, SP4, NR3C1
MIR-380-3P	3/103	0.0317	6.27	SNAP25, BDNF, SP4

Table 8. miRNA for the BPA bi-interacted genes in neurodevelopmental disorders.

6. Comparable chemicals

The CTD provides a way to group chemicals based upon their biological effects, instead of their physical or structural properties, which provides a novel way to view and classify genes and chemicals and will help advance testable hypotheses about environmental chemical-gene disease networks [61]. Comparable chemicals were curated for the possible sharing with many of the networks common to BPA in neurodevelopmental disorders (**Table 9**). Tetrachlorodibenzodioxin, benzo(a)pyrene, vehicle emissions and dibutyle phthalate, as the common environmental pollutants, were found interacting with 312, 269, 204 and 159 of the 403 BPA bi-interacted genes in the NDs, respectively. Drugs such as valproic acid, acetaminophen,

Chemical	CAS RN	Similarity index	CIGs	CIGs for NDs
Tetrachlorodibenzodioxin	1746-01-6	0.5582	12,047	312
Valproic acid	99-66-1	0.5035	10,936	316
Benzo(a)pyrene	50-32-8	0.4511	9511	269
Acetaminophen	103-90-2	0.4093	8435	247
Aflatoxin B1	1162-65-8	0.3851	8140	252
Cyclosporine	59865-13-3	0.3566	7234	187
Nanotubes, carbon		0.3498	7134	199
Vehicle emissions		0.3458	7156	204
Pirinixic acid	50892-23-4	0.3305	6663	201
Estradiol	50-28-2	0.2930	5864	202
Copper Sulfate	7758-98-7	0.2785	5598	145
(6-(4-(2-Piperidin-1-ylethoxy)phenyl))-3-pyridin-4-ylpyrazolo(1,5-a)pyrimidine		0.2583	5186	186
4-(5-Benzo(1,3)dioxol-5-yl-4-pyridin-2-yl-1H-imidazol-2-yl)benzamide		0.2582	5185	183
Ethinyl estradiol	57-63-6	0.2535	5101	153
Tretinoin	302-79-4	0.2428	4835	193
Tetradecanoylphorbol acetate	16561-29-8	0.2315	4530	146
Atrazine	1912-24-9	0.2293	4559	175
Ammonium chloride	12125-02-9	0.2273	4509	215
Dibutyl phthalate	84-74-2	0.2262	4427	159
Silicon dioxide	7631-86-9	0.2224	4441	157

CIGs: common interacting genes; NDs: neurodevelopmental disorders.

Table 9. Chemicals having comparable sets of interacting genes to bisphenol A.

cyclosporine, pirinixic acid, tretinoin and tetradecanoylphorbol Acetate were found interacted with 316, 269, 247, 187, 201, 193 and 146 of the 403 BPA bi-interacted genes, respectively. Dietary pollutant aflatoxin B1, pesticide atrazine, and occupational exposure like copper sulphate, ammonium chloride and silicon dioxide and even estrogen estradiol could interact with the genes of those BPA bi-interacted within the NDs.

7. Future trends and conclusion

With the existed data libraries (mainly CTD, GO, pathway, TFs and miRNA relate databases), bioinformatics softwares (Cytoscape, MCODE and Genemania) or web-based tools (STRING, GEO, ArrayExpress, David and EnrichR), BPs, CCs, MFs, signal pathways and gene regulation in the BPA-gene-disease networks were presented. These data integration and curation yielded insight into the actions of BPA and provide a basis for developing hypotheses about the molecular mechanisms underlying the aetiology of the neurodevelopmental disorder ID, LD, Sch, ASD, AD and BD, although most of the other neurodevelopmental disorders showed no enough information to make a conclusion. The nervous system–related CCs such as neuron related, synapse related, dendrite and axon related are common in CC annotation; the commonly found MFs are neurotransmitter receptor binding or activity, signal transducer or receptor binding or activity; and the main commonly involved BPs include synaptic signalling, cognition, learning or memory, behaviour, the development of nervous system and brain, and the regulation of the related BPs. Neuroactive ligand-receptor interaction, dopaminergic, glutamatergic and serotonergic synapse, monoamine transport, synaptic vesicle pathway may involve in the action of BPA in the neurodevelopmental disorders. Simultaneously, the BPA disease may share the common pathways with drug addictions (cocaine addiction, nicotine addiction and alcoholism), or other types of neurological diseases (Alzheimer's disease, Rett syndrome and sudden infant death syndrome). Unique pathways might also contribute to the BPA action in different NDs like one carbon metabolism and detoxification of oxidative stress–related pathways in Down syndrome. Although GO and pathway results indicate some common characteristics, the predicted PPI molecular function clusters are quite different for each ND. In addition, some of the NDs share the same TFs and miRNAs, which indicate these disorders have the similar expression profiles. What needs to be emphasized that the BPA-gene-disease networks might be influenced by some of the comparable chemicals such as environmental pollutants, drugs, dietary pollutants or occupational exposure, which share the same interacted genes with BPA.

The integrated and curated biological processes and pathways shall shed light on the future studies to find the possible BPA interacted or influenced genes. This will contribute to complete the BPA-disease networks, which surely help to screen the potential biomarker of BPA-induced neurodevelopmental diseases. However, it should be noted that most of the evidences were from curation of the cell or animal experiments. Simultaneously, the biinteraction mode for BPA-gene interaction was adopted for the precise network. Therefore, the future study design should consider the human subjects. Given the sample shortage, the peripheral blood instead of the brain tissue should be preferred in the future. This will contribute to the clinical diagnosis or intervention. Finally, our results should be carefully interpreted because the results might be changed with the increasing abundance of the enrichment of BPA bi-interacted genes.

Author details

Bingling Wang[1*], Ruqin Gao[1] and Da-Hong Wang[2]

*Address all correspondence to: binglingw@yeah.net

1 Qingdao Centers for Disease Control and Prevention, Qingdao, China

2 Department of Biochemistry, Okayama University of Science, Okayama, Japan

References

[1] Rice D, Barone S, Jr. Critical periods of vulnerability for the developing nervous system: Evidence from humans and animal models. Environmental Health Perspectives. 2000;**108(Suppl 3)**:511–533.

[2] Mendonca K, Hauser R, Calafat AM, Arbuckle TE, Duty SM. Bisphenol A concentrations in maternal breast milk and infant urine. International Archives of Occupational and Environmental Health. 2014;**87**:13–20. DOI: 10.1007/s00420-012-0834-9

[3] Ye X, Kuklenyik Z, Needham LL, Calafat AM. Measuring environmental phenols and chlorinated organic chemicals in breast milk using automated on-line column-switching-high performance liquid chromatography-isotope dilution tandem mass spectrometry. Journal of Chromatography B: Analytical Technologies in the Biomedical and Life Sciences. 2006;**831**:110–115. DOI: 10.1016/j.jchromb.2005.11.050

[4] World Health Organization. Joint FAO/WHO Expert Meeting to Review Toxicological and Health Aspects of Bisphenol A. Geneva: World Health Organization; 2010 Nov 1.

[5] Chapin RE, Adams J, Boekelheide K, Gray LE, Hayward SW, Lees PS, McIntyre BS, Portier KM, Schnorr TM, Selevan SG. NTP—CERHR expert panel report on the reproductive and developmental toxicity of bisphenol A. Birth Defects Research Part B: Developmental and Reproductive Toxicology. 2008;**83**:157–395. DOI: 10.1002/bdrb.20147

[6] Healy BF, English KR, Jagals P, Sly PD. Bisphenol A exposure pathways in early childhood: Reviewing the need for improved risk assessment models. Journal of Exposure Science and Environmental Epidemiology. 2015;**25**:544–556. DOI: 10.1038/jes.2015.49

[7] Pinson A, Bourguignon JP, Parent AS. Exposure to endocrine disrupting chemicals and neurodevelopmental alterations. Andrology. 2016;**4**:706–722. DOI: 10.1111/andr.12211

[8] Grondin CJ, Davis AP, Wiegers TC, King BL, Wiegers JA, Reif DM, Hoppin JA, Mattingly CJ. Advancing exposure science through chemical data curation and integration in the comparative toxicogenomics database. Environmental Health Perspectives. 2016; **124**:1592–1599. DOI: 10.1289/EHP174

[9] King BL, Davis AP, Rosenstein MC, Wiegers TC, Mattingly CJ. Ranking transitive chemical-disease inferences using local network topology in the comparative toxicogenomics database. PLoS One. 2012;**7**:e46524. DOI: 10.1371/journal.pone.0046524

[10] Clelland CL, Read LL, Panek LJ, Nadrich RH, Bancroft C, Clelland JD. Utilization of never-medicated bipolar disorder patients towards development and validation of a peripheral biomarker profile. PLoS One. 2013;**8**:e69082. DOI: 10.1371/journal.pone.0069082

[11] Huang da W, Sherman BT, Lempicki RA. Systematic and integrative analysis of large gene lists using DAVID bioinformatics resources. Nature Protocols. 2009;**4**:44–57. DOI: 10.1038/nprot.2008.211

[12] Huang da W, Sherman BT, Lempicki RA. Bioinformatics enrichment tools: Paths toward the comprehensive functional analysis of large gene lists. Nucleic Acids Research. 2009;**37**:1–13. DOI: 10.1093/nar/gkn923

[13] Chen EY, Tan CM, Kou Y, Duan Q, Wang Z, Meirelles GV, Clark NR, Ma'ayan A. Enrichr: Interactive and collaborative HTML5 gene list enrichment analysis tool. BMC Bioinformatics. 2013;**14**:128. DOI: 10.1186/1471-2105-14-128

[14] Kuleshov MV, Jones MR, Rouillard AD, Fernandez NF, Duan Q, Wang Z, Koplev S, Jenkins SL, Jagodnik KM, Lachmann A, McDermott MG, Monteiro CD, Gundersen GW, Ma'ayan A. Enrichr: A comprehensive gene set enrichment analysis web server 2016 update. Nucleic Acids Research. 2016;**44**:W90–97. DOI: 10.1093/nar/gkw377

[15] von Mering C, Huynen M, Jaeggi D, Schmidt S, Bork P, Snel B. STRING: A database of predicted functional associations between proteins. Nucleic Acids Research. 2003;**31**:258–261. DOI: 10.1093/nar/gkg034

[16] Bader GD, Hogue CW. An automated method for finding molecular complexes in large protein interaction networks. BMC Bioinformatics. 2003;**4**:2. DOI: 10.1186/1471-2105-4-2

[17] Shannon P, Markiel A, Ozier O, Baliga NS, Wang JT, Ramage D, Amin N, Schwikowski B, Ideker T. Cytoscape: A software environment for integrated models of biomolecular interaction networks. Genome Research. 2003;**13**:2498–2504. DOI: 10.1101/gr.1239303

[18] Montojo J, Zuberi K, Rodriguez H, Kazi F, Wright G, Donaldson SL, Morris Q, Bader GD. GeneMANIA Cytoscape plugin: Fast gene function predictions on the desktop. Bioinformatics. 2010;**26**:2927–2928. DOI: 10.1093/bioinformatics/btq562

[19] Kochinke K, Zweier C, Nijhof B, Fenckova M, Cizek P, Honti F, Keerthikumar S, Oortveld MA, Kleefstra T, Kramer JM, Webber C, Huynen MA, Schenck A. Systematic phenomics analysis deconvolutes genes mutated in intellectual disability into biologically coherent modules. American Journal of Human Genetics. 2016;**98**:149–164. DOI: 10.1016/j.ajhg.2015.11.024

[20] Kristensen AR, Gsponer J, Foster LJ. A high-throughput approach for measuring temporal changes in the interactome. Nature Methods. 2012;**9**:907–909. DOI: 10.1038/nmeth.2131

[21] Havugimana PC, Hart GT, Nepusz T, Yang H, Turinsky AL, Li Z, Wang PI, Boutz DR, Fong V, Phanse S, Babu M, Craig SA, Hu P, Wan C, Vlasblom J, Dar VU, Bezginov A, Clark GW, Wu GC, Wodak SJ, Tillier ER, Paccanaro A, Marcotte EM, Emili A. A census of human soluble protein complexes. Cell. 2012;**150**:1068–1081. DOI: 10.1016/j.cell.2012.08.011

[22] Hutchins JR, Toyoda Y, Hegemann B, Poser I, Heriche JK, Sykora MM, Augsburg M, Hudecz O, Buschhorn BA, Bulkescher J, Conrad C, Comartin D, Schleiffer A, Sarov M, Pozniakovsky A, Slabicki MM, Schloissnig S, Steinmacher I, Leuschner M, Ssykor A, Lawo S, Pelletier L, Stark H, Nasmyth K, Ellenberg J, Durbin R, Buchholz F, Mechtler K, Hyman AA, Peters JM. Systematic analysis of human protein complexes identifies chromosome segregation proteins. Science. 2010;**328**:593–599. DOI: 10.1126/science.1181348

[23] Stuart JM, Segal E, Koller D, Kim SK. A gene-coexpression network for global discovery of conserved genetic modules. Science. 2003;**302**:249–255. DOI: 10.1126/science.1087447

[24] von Mering C, Krause R, Snel B, Cornell M, Oliver SG, Fields S, Bork P. Comparative assessment of large-scale data sets of protein-protein interactions. Nature. 2002;**417**:399–403. DOI: 10.1038/nature750

[25] Boccuto L, Chen CF, Pittman AR, Skinner CD, McCartney HJ, Jones K, Bochner BR, Stevenson RE, Schwartz CE. Decreased tryptophan metabolism in patients with autism spectrum disorders. Molecular Autism. 2013;**4**:16. DOI: 10.1186/2040-2392-4-16

[26] Rieger KE, Hong WJ, Tusher VG, Tang J, Tibshirani R, Chu G. Toxicity from radiation therapy associated with abnormal transcriptional responses to DNA damage. Proceedings of the National Academy of Sciences of the United States of America. 2004;**101**:6635–6640. DOI: 10.1073/pnas.0307761101

[27] Varjosalo M, Sacco R, Stukalov A, van Drogen A, Planyavsky M, Hauri S, Aebersold R, Bennett KL, Colinge J, Gstaiger M, Superti-Furga G. Interlaboratory reproducibility of large-scale human protein-complex analysis by standardized AP-MS. Nature Methods. 2013;**10**:307–314. DOI: 10.1038/nmeth.2400

[28] Varjosalo M, Keskitalo S, Van Drogen A, Nurkkala H, Vichalkovski A, Aebersold R, Gstaiger M. The protein interaction landscape of the human CMGC kinase group. Cell Reports. 2013;**3**:1306–1320. DOI: 10.1016/j.celrep.2013.03.027

[29] Jeronimo C, Forget D, Bouchard A, Li Q, Chua G, Poitras C, Therien C, Bergeron D, Bourassa S, Greenblatt J, Chabot B, Poirier GG, Hughes TR, Blanchette M, Price DH, Coulombe B. Systematic analysis of the protein interaction network for the human transcription machinery reveals the identity of the 7SK capping enzyme. Molecular Cell. 2007;**27**:262–274. DOI: 10.1016/j.molcel.2007.06.027

[30] Hashimoto S, Boissel S, Zarhrate M, Rio M, Munnich A, Egly JM, Colleaux L. MED23 mutation links intellectual disability to dysregulation of immediate early gene expression. Science. 2011;**333**:1161–1163. DOI: 10.1126/science.1206638

[31] Melko M, Douguet D, Bensaid M, Zongaro S, Verheggen C, Gecz J, Bardoni B. Functional characterization of the AFF (AF4/FMR2) family of RNA-binding proteins: Insights into the molecular pathology of FRAXE intellectual disability. Human Molecular Genetics. 2011;**20**:1873–1885. DOI: 10.1093/hmg/ddr069

[32] Richard DJ, Cubeddu L, Urquhart AJ, Bain A, Bolderson E, Menon D, White MF, Khanna KK. hSSB1 interacts directly with the MRN complex stimulating its recruitment to DNA double-strand breaks and its endo-nuclease activity. Nucleic Acids Research. 2011;**39**:3643–3651. DOI: 10.1093/nar/gkq1340

[33] Baynam G, Overkov A, Davis M, Mina K, Schofield L, Allcock R, Laing N, Cook M, Dawkins H, Goldblatt J. A germline MTOR mutation in Aboriginal Australian siblings with intellectual disability, dysmorphism, macrocephaly, and small thoraces. American Journal of Medical Genetics A. 2015;**167**:1659–1667. DOI: 10.1002/ajmg.a.37070

[34] Perluigi M, Pupo G, Tramutola A, Cini C, Coccia R, Barone E, Head E, Butterfield DA, Di Domenico F. Neuropathological role of PI3K/Akt/mTOR axis in Down syndrome brain. Biochimica et Biophysica Acta. 2014;**1842**:1144–1153. DOI: 10.1016/j.bbadis.2014.04.007

[35] Troca-Marin JA, Alves-Sampaio A, Montesinos ML. Deregulated mTOR-mediated translation in intellectual disability. Progress in Neurobiology. 2012;**96**:268–282. DOI: 10.1016/j.pneurobio.2012.01.005

[36] Xiang HG, Hao J, Zhang WJ, Lu WJ, Dong P, Liu YB, Chen L. Expression of fatty acid synthase negatively correlates with PTEN and predicts peritoneal dissemination of human gastric cancer. Asian Pacific Journal of Cancer Prevention. 2015;**16**:6851–6855. DOI: 10.7314/APJCP.2015.16.16.6851

[37] Boldrick JC, Alizadeh AA, Diehn M, Dudoit S, Liu CL, Belcher CE, Botstein D, Staudt LM, Brown PO, Relman DA. Stereotyped and specific gene expression programs in human innate immune responses to bacteria. Proceedings of the National Academy of Sciences of the United States of America. 2002;**99**:972–977. DOI: 10.1073/pnas.231625398

[38] Mariani M, Crosti F, Redaelli S, Fossati C, Piras R, Biondi A, Dalpra L, Selicorni A. Partial duplication of the PARK2 gene in a child with developmental delay and her normal mother: A second report. American Journal of Medical Genetics B: Neuropsychiatric Genetics. 2013;**162B**:485–486. DOI: 10.1002/ajmg.b.32173

[39] Scheuerle A, Wilson K. PARK2 copy number aberrations in two children presenting with autism spectrum disorder: Further support of an association and possible evidence for a new microdeletion/microduplication syndrome. American Journal of Medical Genetics B: Neuropsychiatric Genetics. 2011;**156B**:413–420. DOI: 10.1002/ajmg.b.31176

[40] Wu G, Feng X, Stein L. A human functional protein interaction network and its application to cancer data analysis. Genome Biology. 2010;**11**:R53. DOI: 10.1186/gb-2010-11-5-r53

[41] Gysin S, Paquette J, McMahon M. Analysis of mRNA profiles after MEK1/2 inhibition in human pancreatic cancer cell lines reveals pathways involved in drug sensitivity. Molecular Cancer Research. 2012;**10**:1607–1619. DOI: 10.1158/1541-7786.MCR-12-0188

[42] Ramaswamy S, Tamayo P, Rifkin R, Mukherjee S, Yeang C-H, Angelo M, Ladd C, Reich M, Latulippe E, Mesirov JP. Multiclass cancer diagnosis using tumor gene expression signatures. Proceedings of the National Academy of Sciences of the United States of America. 2001;**98**:15149–15154. DOI: 10.1073/pnas.211566398

[43] Kang H, Chen IM, Wilson CS, Bedrick EJ, Harvey RC, Atlas SR, Devidas M, Mullighan CG, Wang X, Murphy M, Ar K, Wharton W, Borowitz MJ, Bowman WP, Bhojwani D, Carroll WL, Camitta BM, Reaman GH, Smith MA, Downing JR, Hunger SP, Willman CL. Gene expression classifiers for relapse-free survival and minimal residual disease improve risk classification and outcome prediction in pediatric B-precursor acute lymphoblastic leukemia. Blood. 2010;**115**:1394–1405. DOI: 10.1182/blood-2009-05-218560

[44] Bild AH, Yao G, Chang JT, Wang Q, Potti A, Chasse D, Joshi MB, Harpole D, Lancaster JM, Berchuck A, Olson JA, Jr., Marks JR, Dressman HK, West M, Nevins JR. Oncogenic pathway

signatures in human cancers as a guide to targeted therapies. Nature. 2006;**439**:353–357. DOI: 10.1038/nature04296

[45] Lin A, Wang RT, Ahn S, Park CC, Smith DJ. A genome-wide map of human genetic interactions inferred from radiation hybrid genotypes. Genome Research. 2010;**20**:1122–1132. DOI: 10.1101/gr.104216.109

[46] Lee YH, Kim JH, Song GG. Pathway analysis of a genome-wide association study in schizophrenia. Gene. 2013;**525**:107–115. DOI: 10.1016/j.gene.2013.04.014

[47] Emamian ES. AKT/GSK3 signaling pathway and schizophrenia. Frontiers in Molecular Neuroscience. 2012;**5**:33. DOI: 10.3389/fnmol.2012.00033

[48] Arijs I, De Hertogh G, Lemaire K, Quintens R, Van Lommel L, Van Steen K, Leemans P, Cleynen I, Van Assche G, Vermeire S, Geboes K, Schuit F, Rutgeerts P. Mucosal gene expression of antimicrobial peptides in inflammatory bowel disease before and after first infliximab treatment. PLoS One. 2009;**4**:e7984. DOI: 10.1371/journal.pone.0007984

[49] Anderson OS, Nahar MS, Faulk C, Jones TR, Liao C, Kannan K, Weinhouse C, Rozek LS, Dolinoy DC. Epigenetic responses following maternal dietary exposure to physiologically relevant levels of bisphenol A. Environmental and Molecular Mutagenesis. 2012;**53**:334–342. DOI: 10.1002/em.21692

[50] Dolinoy DC, Huang D, Jirtle RL. Maternal nutrient supplementation counteracts bisphenol A-induced DNA hypomethylation in early development. Proceedings of the National Academy of Sciences of the United States of America. 2007;**104**:13056–13061. DOI: 10.1073/pnas.0703739104

[51] Coskun PE, Busciglio J. Oxidative stress and mitochondrial dysfunction in Down's syndrome: Relevance to aging and dementia. Current Gerontology and Geriatrics Research. 2012;**2012**:383170. DOI: 10.1155/2012/383170

[52] Infantino V, Castegna A, Iacobazzi F, Spera I, Scala I, Andria G, Iacobazzi V. Impairment of methyl cycle affects mitochondrial methyl availability and glutathione level in Down's syndrome. Molecular Genetics and Metabolism. 2011;**102**:378–382. DOI: 10.1016/j.ymgme.2010.11.166

[53] Hobert O. Common logic of transcription factor and microRNA action. Trends in Biochemical Sciences. 2004;**29**:462–468. DOI: 10.1016/j.tibs.2004.07.001

[54] Hobert O. Gene regulation by transcription factors and microRNAs. Science. 2008;**319**:1785–1786. DOI: 10.1126/science.1151651

[55] Chandler SP, Kansagra P, Hirst MC. Fragile X (CGG) n repeats induce a transcriptional repression in cis upon a linked promoter: Evidence for a chromatin mediated effect. BMC Molecular Biology. 2003;**4**:3. DOI: 10.1186/1471-2199-4-3

[56] Kumari D, Gabrielian A, Wheeler D, Usdin K. The roles of Sp1, Sp3, USF1/USF2 and NRF-1 in the regulation and three-dimensional structure of the Fragile X mental retardation gene promoter. Biochemical Journal. 2005;**386**:297–303. DOI: 10.1042/BJ20041124

[57] Ng SY, Johnson R, Stanton LW. Human long non-coding RNAs promote pluripotency and neuronal differentiation by association with chromatin modifiers and transcription factors. EMBO Journal. 2012;**31**:522–533. DOI: 10.1038/emboj.2011.459

[58] Chong JA, Tapia-Ramirez J, Kim S, Toledo-Aral JJ, Zheng Y, Boutros MC, Altshuller YM, Frohman MA, Kraner SD, Mandel G. REST: A mammalian silencer protein that restricts sodium channel gene expression to neurons. Cell. 1995;**80**:949–957. DOI: 10.1016/0092-8674(95)90298-8

[59] Potkin SG, Turner J, Fallon J, Lakatos A, Keator D, Guffanti G, Macciardi F. Gene discovery through imaging genetics: Identification of two novel genes associated with schizophrenia. Molecular Psychiatry. 2009;**14**:416–428. DOI: 10.1038/mp.2008.127

[60] Muinos-Gimeno M, Guidi M, Kagerbauer B, Martin-Santos R, Navines R, Alonso P, Menchon JM, Gratacos M, Estivill X, Espinosa-Parrilla Y. Allele variants in functional MicroRNA target sites of the neurotrophin-3 receptor gene (NTRK3) as susceptibility factors for anxiety disorders. Human Mutatation. 2009;**30**:1062–1071. DOI: 10.1002/humu.21005

[61] Davis AP, Murphy CG, Saraceni-Richards CA, Rosenstein MC, Wiegers TC, Hampton TH, Mattingly CJ. GeneComps and ChemComps: A new CTD metric to identify genes and chemicals with shared toxicogenomic profiles. Bioinformation. 2009;**4**:173–174. DOI: 10.6026/97320630004173

Section 2

Prenatal and Postnatal Effects of Bisphenol A

Chapter 6

Low-Dose Exposure to Bisphenol A in Early Life

Yeon-Pyo Hong and Yun-Jung Yang

Additional information is available at the end of the chapter

http://dx.doi.org/10.5772/intechopen.68428

Abstract

Bisphenol A (BPA) has lower estrogenic potency than 17b-estadiol. The reference dose of BPA is defined as 50 ug/kg bw/day by the Environmental Protection Agency. The lower doses of BPA than no observable effect level are considered safe. However, early life exposure to low-dose BPA may increase the risk of developing adult onset disease. The harmful effects caused by low-dose BPA in fetus and newborns can transmit to third or fourth generations. The suggested mechanism of transgeneration is epigenetic changes. In addition, simultaneous exposure to various chemicals can induce combined effects. Low-dose effects of BPA are ongoing controversy because the animal test results will be the same in humans. Epidemiologic evidences are needed to provide the human health effects from exposure to low dose of BPA.

Keywords: bisphenol A, low dose, early life

1. Introduction

Bisphenol A (BPA, CAS #80-05-7) is widely used in manufacturing polycarbonate plastics, epoxy resins, and thermal paper including food containers, baby bottles, dental sealant, and store receipts. BPA can produce the estrogenic activity through binding with estrogen receptor [1] and can also exert the actions through androgen receptor, peroxisome proliferator-activated receptor γ, and others [2]. Thus, BPA is classified as an endocrine disruptor (ED) because of the estrogenic potency.

Traditional toxicology considers that the effect would consistently increase with the amount of treatment. The dose levels of BPA below low observed adverse effects level (LOAEL) at 50 mg/kg/day [3] are regard as safe. However, recent studies describe the nonmonotonic dose response relationship of BPA. Perinatal exposure to lower dose of BPA than LOAEL has been reported the harmful effects on endocrine system including reproductive system [4–7], immune system [8–10], pituitary gland [11–20], and metabolic system [21–24]. Because, the fetus and neonates are extremely sensitive to perturbation by hormone like chemicals, early life exposure to low dose BPA probably is able to affect the epigenetic mechanism. The epigenetic changes caused by BPA may explain the increased risk of developing adult onset diseases.

The mixed exposure to several low-level EDs should be tested because humans are exposed to various EDs simultaneously. Mixture can produce significant adverse effects, even when each chemical is present at low doses that individually do not induce observable effects in reproductive system [4, 25] neuroendocrine system [26], and endocrine system [27, 28].

Therefore, this chapter describes the harmful effects on adulthood caused by exposure to low-dose BPA during early life stage.

2. Exposure to bisphenol A

BPA is a chemical compound used to produce polycarbonate plastic and epoxy resins. Therefore, humans are exposed to BPA throughout their life. The predominant source of BPA exposure to general population is ingestion of food and beverages [29, 30]. Humans can also be exposed to BPA through nondietary routes including inhalation [31, 32] and skin contact [33].

2.1. Human exposure levels

BPA levels have been measured in human biological samples due to the widespread use of BPA-containing products. Unconjugated BPA levels in humans were measured as a wide range from 0.2 to 20 ng/ml in serum [32, 34]. BPA also detected in amniotic fluid [35], breast milk [36, 37], and maternal amniotic fluid and fetal plasma [38, 39]. These studies indicate that BPA is able to easily transverse the placental barrier and affect the fetal development.

The estimation of BPA exposure in the general population can be based on the presence of BPA levels in the biological sample and the amount of daily food intake. Based on urinary excretion levels of BPA metabolites, the estimated amounts of BPA in general population are up to 0.16-μg/kg body weight (bw) in the USA and 0.04–0.08 μg/kg bw in Japan [40]. Daily intake levels of BPA to human have been estimated from 0.2-μg/kg bw/day in 3-month-old breasted infants up to 13-μg/kg bw/day in 6-to-12-month-old infants. The estimates of potential dietary exposure in young children and adults were respectively 5.3 and 1.5 μg/kg bw/day based on conservative migration values of BPA and conservative estimates of consumption of commercial foods and beverages [41]. This report shows that infants and children are the highest intake group because they eat, drink, and breathe more than adults and play or bite with the plastic toys.

2.2. Metabolism

The orally administered BPA could rapidly metabolize to the bisphenol A-glucuronide carried out by the uridine 5′-diphospho-glucuronyl transferase (UGT) in the liver and gut. The metabolic process is called as glucuronidation. Unconjugated parent BPA is converted into other substances such as sulphate conjugate [37]. The conjugated form of BPA does not bind to the estrogen receptors and is excreted in urine [42–44]. It suggests that the conjugates have relatively less estrogenic potency than unconjugated form because of the less binding affinity to nuclear receptors and rapid excretion. The estimated half-life of BPA was about 6 h in the human body [45–47]. In addition, BPA metabolisms in liver cells from rats, mice, and humans showed similar pattern across the species [48].

The metabolic process of BPA is different depending on the route of exposure. The highest concentration of BPA was measured at 1 h after oral or intraperitoneal administration and at 4 h after subcutaneous administration [44]. More than 60% of the glucuronidated BPA were excreted through urine, and unconjugated form of BPA was mainly excreted in the feces [44]. It suggested that the oral route was recommended for appropriated risk assessment of BPA because the predominant exposure route of BPA to human is dietary ingestion [40].

In pharmacokinetic studies, BPA metabolites were measured in human urine and blood after ingestion of 5 mg deuterated (d6)-BPA [46]. Besides, the maximum unconjugated d6-BPA concentration in human serum was detected at 1.6 h after ingestion of the BPA-contained soup [49].

The fetus and newborns are not fully developed in the ability of glucuronidation [50]. In experimental studies, neonates showed higher free BPA levels in blood compared to older animals when given a same level of BPA [50, 51]. Despite the glucuronidation enzymes have not been identified in human, neonates and infants may be vulnerable to BPA exposure compared to adult human.

3. Health effects of low dose bisphenol A

Until recently, the studies on BPA mainly focused on the nuclear mechanisms of estrogen response through bind with estrogen receptors (ERs). The binding affinity of BPA to ERβ is about 10 times higher than that of ERα [52, 53]. BPA showed 10,000–100,000 weaker estrogenic potencies compared to 17β-estradiol [54]. It has been considered that BPA has relatively weak estrogenic potency due to the low binding affinity with ERs and the low estrogenic potency compared to estradiol.

However, recent studies reported a variety of molecular pathways including androgen receptor, aryl hydrocarbon receptor, and peroxisome proliferator-activated receptor, which are associated with hormones of the endocrine and other systems in the body [34, 55]. The disrupted nuclear hormone receptors can interfere with the secretion and function of endocrine system.

3.1. Low dose effects of bisphenol A

The low dose was defined in the U.S. Environmental Protection Agency (EPA), the National Toxicology Program (NTP) assembled a group of scientists in 2001 as any biological effects occurring in the range of typical human exposures or occurring at doses lower than those typically used in traditional toxicology assessment [56]. Traditional toxicology considers that the dose makes poison. Thus, toxicological studies have been focused on identifying the concentrations at which chemicals can cause biological changes, and below that levels are not harmful to health.

According to the definition of NTP, the cutoff doses of low-dose BPA might be the range of general public exposure except for occupational exposure and the levels less than 50 mg/kg/day of LOAEL [3, 54]. However, diethylstilbestrol (DES), which was used to prevent premature births and miscarriages of pregnant women, is one of the endocrine disruptor and is caused endocrine disrupting activity to exposed women and developing babies [57]. Thus, the safety levels of EDs may not exist.

Many experimental studies have been reported on low dose effects of BPA [1, 58, 59]. Epidemiologic studies also showed that exposure to environmental relevant levels of BPA are associated with the disorders in human [60–62]. However, there is still controversy over the low-dose effects of BPA because of the difficulty to replicate. Thus, the necessity of the reevaluation of human safety daily intake limits is raised.

Low dose is not the same as nonmonotonicity. Monotonic dose response relationship is the basic approach in traditional toxicology. In contrast to traditional toxicological approach, recent studies suggest that EDs may show the nonmonotonicity including biphasic, U- or inverted U-shape dose–response curve (**Figure 1**) [63]. The lack of monotonic dose-response relationship makes it difficult to predict the health effects at low dose using the result from high-dose endocrine disruptors.

Exposure to environmental relevant doses of BPA to pregnant mice moved the timing of vaginal opening and first estrous cyclicity up in their offspring [58]. BPA below reference dose affects the structure and functions of brain through interfering with the hormones and neuro hormone receptors [64]. It may be caused by the disruption on brain-gonads-pituitary

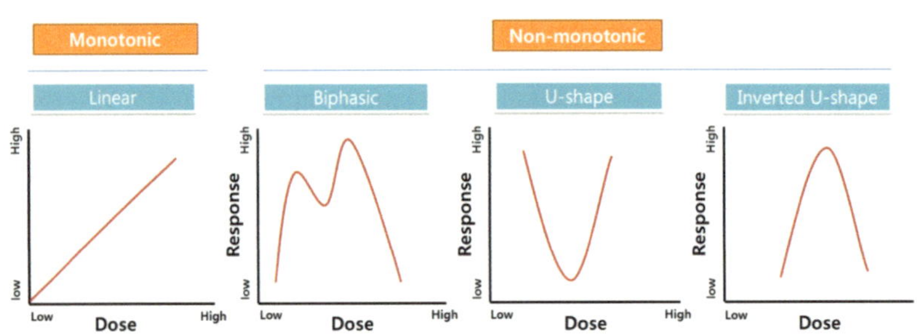

Figure 1. Examples of monotonic and non-monotonic dose response curve.

gland axis function. However, BPA exposed male and female rats showed no changes of body weight, reproductive morphology, and fertility of their female offspring [65].

In epidemiologic studies, the associations were observed between internal BPA concentrations and endocrine hormones. The BPA concentrations in the urine of men in the fertility clinic were showed inverse correlation with the estradiol:testosterone ratio [66]. Urinary BPA concentrations in human from Italy were positively associated with ERα and ERβ [67].

Recent study showed that BPA at low doses decreased estradiol level and inhibited growth of follicles isolated from wild-type and aryl hydrocarbon receptor (AHR) knock-out mice through interfering with the AHR [55]. They suggested that AHR signaling pathway might not be a major route through BPA exert its toxic effect on ovarian follicles.

The low-dose effects of BPA may associate with the genetic susceptibility, i.e., a gene-environment interaction. Transgenerational inheritance may associate with the epigenetic changes caused by low-dose BPA exposure. Without understanding the gene-environment interactions, there is a limit to understand the low dose effects. Low-dose effects of BPA should be validated through epidemiologic studies.

Exposed environmental factors during fetal or neonatal life can interact with the genome and influence the onset of diseases in their adulthood including cancer, infertility, precocious puberty, and obesity [68]. This theory is called "the developmental origins of health and disease" [69]. DES, a synthetic estrogen, is well documented that fetal exposure to DES causes the severe malformations and cancers of the reproductive tract [57].

Perinatal exposure to low-dose BPA may produce the adverse effects including brain function, reproduction, pituitary gland, and immunity (**Table 1**). The harmful effects are persisted and transferred to the fourth generation that was not directly exposed to BPA. BPA exposed fetus during their gestational period showed neoplasia and changes in mammary tissue [70].

Organ developing period as the first trimester in fetus is the critical period, which means they are extremely sensitive to low-dose effects of EDs than adult organisms. Thus, gestational exposure to EDs may induce the harmful effects on the offspring and can transfer to the subsequent generation. This process is called as "epigenetic transgenerational inheritance." The attention has been increasing to the role of epigenetic changes in the development of disease because it is considered as one of the mechanisms for explaining of low-dose effects.

When epigenetic changes are induced by EDs, those can regulate the gene expression by silencing or activating the gene. The mechanisms of regulation are classified as (1) DNA methylation, (2) histone modification, and (3) RNA-associated silencing. Because epigenetic changes do not modify the gene sequence but affect the gene expression, it may reflect the plausible association between exposure to endocrine disruptors and alteration of gene expression, which resulted into the development of disease.

Classification	Animal model	Administration (dose)	Exposure duration	Effects (offspring)	References
Reproductive system	Pregnant rats	Gavage (50 μg/kg bw/day)	GD 6–PND 21	Reduction of semen quality	[4]
Reproductive system	Medaka (Oryzias latipes)	Water tank (200 ng/ml)	Lifelong, development, neurogenesis, sex differentiation	Transgenerational effects	[71]
Reproductive system	Pregnant rats	Drinking water (3 μg/kg bw/day [estimated average dose of exposure])	GD 0–PND 21	Increase LH, estradiol levels in serumAbnormal ovary histology	[5]
Reproductive system	Pregnant rats	Gavage (2.5, 25, 260, 2700 ug/kg bw/day)	GD 6–GD 21 (Dam gavage)/ PND 0–PND 21 (Pup gavage)	Alters estrogen receptor expression	[7]
Pituitary system/Brain	Pregnant rats	Gavage (2.5, 25, 2500 μg/kg bw/day)	GD 6–GD 21 (Dam gavage)/ PND 0–PND 21 (Pup gavage)	Effects on anxiety or exploratory activity No consistent effects	[12]
Pituitary system/Brain	Pregnant rats	Oral (440,400 μg/kg bw/day)	GD 6–PND 21	Alters NCS proliferation and differentiation	[11]
Pituitary system/Brain	Pregnant rats	Oral (40 μg/kg bw/day)	GD 0–PND 21	Cause anxiety like alteration	[13]
Pituitary system/Brain	Pregnant rats	SC injection (10 μg/kg bw/day)	GD 12–PND 21	Disruption in dopamine- and serotonin-related genes	[14]
Pituitary system/Brain	Pregnant rats	Oral (5 μg/kg bw/day)	GD 1–PND 100	Adverse development and behavior effects on F1 and F2	[17]
Reproduction disorders	Pregnant rats	Drinking water (3 μg/kg bw/day [estimated average dose of exposure])	GD 0–PND 21	Increase FSH, LH levels in serum Abnormal testis histology	[6]
Pituitary system/Brain	Pregnant rats	Diet (40 μg/kg bw/day)	GD 0–PND 21	Abnormal adrenal histology Alters the basal and stress induced activity	[15]
Pituitary system/Brain	Pregnant rats	Drinking water (0.1 mg/L BPA)	PND 0–PND 21	Alters the ERα signaling and behavioral deficit	[16]
Pituitary system/Brain	Pregnant rats	Gavage (10, 100, 1000, 10,000 μg/kg bw/day)	GD 10–PND 10	Alters brain development	[18]

http://dx.doi.org/10.5772/intechopen.68428

Classification	Animal model	Administration (dose)	Exposure duration	Effects (offspring)	References
Pituitary system/Brain	Pregnant rats	Gavage (5, 10 μg/kg bw/day)	GD 6–GD 21	Alters brain development	[19]
Pituitary system/Brain	Four week old mice (male and female)	Gavage (40, 400 μg/kg bw/day)	8 weeks	Sex difference of behaviors: locomotion, exploration, anxiety, learning-memory in adult	[20]
Immune system	Pregnant rats	Gavage (5 μg/kg bw/day)	GD 15–PND 21	Alters immune response	[8]
Immune system	Pregnant rats	Gavage (0.5, 5, 50, 500 μg/kg bw/day)	GD 6–PND 21	Induce allergic inflammation	[9]
Immune system	Pregnant mice	Gavage (50 μg/kg bw/day)	GD 6–PND 21	Affect immune response	[72]
Immune system	Pregnant mice	Drinking water (10 μg/ml)	GD −28 (One week before mating)–PND 22	increase the development risk of asthma	[10]
Metabolic system	Female mice (4-week-old)	Diet (50 μg/kg bw/day)	Before mating–7 weeks after birth	Alters metabolism	[21]
Metabolic system	Pregnant mice	Diet (Prenatal: 0.19, 3.49 μg/kg bw/day Postnatal: 0.36, 7.2 μg/kg bw/day)	GD 0–PND 21	Control food intake and energy expenditure	[22]
Metabolic system	Pregnant rats	Drinking water (0.01, 0.1, 1.0 C31 mg/L)	GD 11–PND 21	Increase the preference for sweet taste Sexual difference (female>male)	[23]
Metabolic system	Pregnant rats	Drinking water (1 mg/L BPA (estimated 70 μg/kg bw/day))	GD 6–PND 21	Increase adipogenesis: gene expression	[24]

Table 1. Low-dose studies of BPA in early life stage.

3.2. Mixed exposure

Traditional risk assessment approaches are focused on the single chemical. Individual NOAEL does not reveal about the possible risk for the multiple exposure of EDs. Simultaneous exposure to multiple endocrine disruptors (mixed exposure) can generate combination effects even lower than their NOAEL [73, 74].

The crucial definitions for assessment of mixture exposure are classified as synergisms, antagonisms, or additivity: synergism means that the observed effects are stronger than expected; likewise, if they are weaker than expectations, there is antagonism. The combination effects are similar to the effect of individual agents are called additivism [75].

Combined effects have been reported when treated with mixture of BPA and other EDs simultaneously. *In vitro* studies, synergistic/additive effects are showed in case of simultaneous exposure of two or more chemicals [76, 77]. Perinatal exposure to low dose of BPA and paraben showed the additive effects on the downregulated semen quality in adult male offspring compared to individual exposure [4]. Mixture of BPA and other plastic-derived chemicals, despite of higher dose than environmental relevant levels, promoted epigenetic transgenerational inheritance of adult onset disease including obesity, testis, and ovary disease [78].

4. Conclusion

BPA in daily life are considered safe; however, low-dose effects are observed in experimental studies. Early life exposure to low-dose BPA may increase the risk of developing adult onset of disease, and the biological changes can transmit to the third or fourth generation. Therefore, EFSA propose the tolerable daily intake levels of BPA from 50 to 4-ug/kg bw/day. Low-dose effects of BPA are ongoing controversy because of the inconsistent results. Epidemiologic evidences such as nested case control studies are needed to provide the human health effects caused by exposure to low dose of BPA.

Author details

Yeon-Pyo Hong[1]* and Yun-Jung Yang[2]

*Address all correspondence to: hyp026@cau.ac.kr

1 Department of Preventive Medicine, College of Medicine, Chung-Ang University, Seoul, South Korea

2 Institute for Integrative Medicine, International St. Mary's Hospital, Catholic Kwandong University, Gangneung, South Korea

References

[1] Vandenberg LN, Colborn T, Hayes TB, Heindel JJ, Jacobs DR Jr, Lee DH, Shioda T, Soto AM, vom Saal FS, Welshons WV, Zoeller RT, Myers JP. Hormones and endocrine-disrupting chemicals: Low-dose effects and nonmonotonic dose responses. Endocrine Reviews. 2012;**33**(3):378-455. Review

[2] Watson CS, Jeng YJ, Guptarak J. Endocrine disruption via estrogen receptors that participate in nongenomic signaling pathways. The Journal of Steroid Biochemistry and Molecular Biology. 2011;**127**(1–2):44-50

[3] US Environmental Protection Agency. Bisphenol A Action Plan [Internet]. 2010:3. Available from: https://www.epa.gov/sites/production/files/2015-09/documents/bpa_action_plan.pdf

[4] Yang YJ, Hong YP, Chae SA. Reduction in semen quality after mixed exposure to bisphenol A and isobutylparaben in utero and during lactation periods. Human & Experimental Toxicology. 2016 Aug;**35**(8):902-911

[5] Gámez JM, Penalba R, Cardoso N, Bernasconi PS, Carbone S, Ponzo O, Pandolfi M, Scacchi P, Reynoso R. Exposure to a low dose of bisphenol A impairs pituitary-ovarian axis in prepubertal rats: Effects on early folliculogenesis. Environmental Toxicology & Pharmacology. 2015;**39**(1):9-15

[6] Gámez JM, Penalba R, Cardoso N, Ponzo O, Carbone S, Pandolfi M, Scacchi P, Reynoso R. Low dose of bisphenol A impairs the reproductive axis of prepuberal male rats. Journal of Physiology and Biochemistry. 2014 Mar;**70**(1):239-246

[7] Rebuli ME, Cao J, Sluzas E, Delclos KB, Camacho L, Lewis SM, Vanlandingham MM, Patisaul HB.. Investigation of the effects of subchronic low dose oral exposure to bisphenol A (BPA) and ethinyl estradiol (EE) on estrogen receptor expression in the juvenile and adult female rat hypothalamus. Toxicological Sciences. 2014;**140**(1):190-203

[8] Ménard S, Guzylack-Piriou L, Lencina C, Leveque M, Naturel M, Sekkal S, Harkat C, Gaultier E, Olier M, Garcia-Villar R, Theodorou V, Houdeau E. Perinatal exposure to a low dose of bisphenol A impaired systemic cellular immune response and predisposes young rats to intestinal parasitic infection. PLoS One. 2014;**9**(11):e112752

[9] Bauer SM, Roy A, Emo J, Chapman TJ, Georas SN, Lawrence BP. The effects of maternal exposure to bisphenol A on allergic lung inflammation into adulthood. Toxicological Sciences. 2012;**130**(1):82-93

[10] Nakajima Y, Goldblum RM, Midoro-Horiuti T. Fetal exposure to bisphenol A as a risk factor for the development of childhood asthma: an animal model study. Environmental Health. 2012;**21**(11):8

[11] Tiwari SK, Agarwal S, Seth B, Yadav A, Ray RS, Mishra VN, Chaturvedi RK.. Inhibitory effects of Bisphenol-A on neural stem cells proliferation and differentiation

in the rat brain are dependent on Wnt/β-Catenin pathway. Molecular Neurobiology. 2015;**52**(3):1735-1757

[12] Rebuli ME, Camacho L, Adonay ME, Reif DM, Aylor DL, Patisaul HB. Impact of low-dose oral exposure to Bisphenol A (BPA) on juvenile and adult rat exploratory and anxiety behavior: A CLARITY-BPA Consortium Study. Toxicological Sciences. 2015;**148**(2):341-354

[13] Zhou R, Chen F, Feng X, Zhou L, Li Y, Chen L. Perinatal exposure to low-dose of bisphenol A causes anxiety-like alteration in adrenal axis regulation and behaviors of rat offspring: A potential role for metabotropic glutamate 2/3 receptors. Journal of Psychiatric Research. 2015;**64**:121-129

[14] Castro B, Sánchez P, Miranda MT, Torres JM, Ortega E. Identification of dopamine- and serotonin-related genes modulated by bisphenol A in the prefrontal cortex of male rats. Chemosphere. 2015 Nov;**139**:235-239

[15] Panagiotidou E, Zerva S, Mitsiou DJ, Alexis MN, Kitraki E.. Perinatal exposure to low-dose bisphenol A affects the neuroendocrine stress response in rats. The Journal of Endocrinology. 2014;**220**(3):207-218

[16] Xu XB, He Y, Song C, Ke X, Fan SJ, Peng WJ, Tan R, Kawata M, Matsuda K, Pan BX, Kato N. Bisphenol A regulates the estrogen receptor alpha signaling in developing hippocampus of male rats through estrogen receptor. Hippocampus. 2014;**24**(12):1570-1580

[17] Boudalia S, Berges R, Chabanet C, Folia M, Decocq L, Pasquis B, Abdennebi-Najar L, Canivenc-Lavier MC. A multi-generational study on low-dose BPA exposure in Wistar rats: Effects on maternal behavior, flavor intake and development. Neurotoxicology and Teratology. 2014;**41**:16-26

[18] McCaffrey KA, Jones B, Mabrey N, Weiss B, Swan SH, Patisaul HB. Sex specific impact of perinatal bisphenol A (BPA) exposure over a range of orally administered doses on rat hypothalamic sexual differentiation. Neurotoxicology. 2013;**36**:55-62

[19] Cao J, Rebuli ME, Rogers J, Todd KL, Leyrer SM, Ferguson SA, Patisaul HB. Prenatal bisphenol A exposure alters sex-specific estrogen receptor expression in the neonatal rat hypothalamus and amygdala. Toxicological Sciences. 2013;**133**(1):157-173

[20] Xu X, Tian D, Hong X, Chen L, Xie L. Sex-specific influence of exposure to bisphenol-A between adolescence and young adulthood on mouse behaviors. Neuropharmacology. 2011;**61**(4):565-573

[21] Naville D, Labaronne E, Vega N, Pinteur C, Canet-Soulas E, Vidal H, Le Magueresse-Battistoni B. Metabolic outcome of female mice exposed to a mixture of low-dose pollutants in a diet-induced obesity model. PLoS One. 2015;**10**(4):e0124015

[22] Mackay H, Patterson ZR, Khazall R, Patel S, Tsirlin D, Abizaid A. Organizational effects of perinatal exposure to bisphenol-A and diethylstilbestrol on arcuate nucleus

circuitry controlling food intake and energy expenditure in male and female CD-1 mice. Endocrinology. 2013;**154**(4):1465-1475

[23] Xu X, Tan L, Himi T, Sadamatsu M, Tsutsumi S, Akaike M, Kato N. Changed preference for sweet taste in adulthood induced by perinatal exposure to bisphenol A-A probable link to overweight and obesity. Neurotoxicology and Teratology. 2011;**33**(4):458-463

[24] Somm E, Schwitzgebel VM, Toulotte A, Cederroth CR, Combescure C, Nef S, Aubert ML, Hüppi PS. Perinatal exposure to bisphenol a alters early adipogenesis in the rat. Environmental Health Perspectives. 2009;**117**(10):1549-1555

[25] Li D, Hu Y, Shen X, Dai X, Han X. Combined effects of two environmental endocrine disruptors nonyl phenol and di-n-butyl phthalate on rat Sertoli cells *in vitro*. Reproductive Toxicology. 2010;**30**(3):438-445

[26] Khurana S, Ranmal S, Ben-Jonathan N. Exposure of newborn male and female rats to environmental estrogens: Delayed and sustained hyperprolactinemia and alterations in estrogen receptor expression. Endocrinology. 2000;**141**(12):4512-4517

[27] Jones S, Boisvert A, Naghi A, Hullin-Matsuda F, Greimel P, Kobayashi T, Papadopoulos V, Culty M. Stimulatory effects of combined endocrine disruptors on MA-10 Leydig cell steroid production and lipid homeostasis. Toxicology. 2016;**355–356**:21-30

[28] Biemann R, Fischer B, Navarrete Santos A. Adipogenic effects of a combination of the endocrine-disrupting compounds bisphenol A, diethylhexylphthalate, and tributyltin. Obesity and the Facts. 2014;**7**(1):48-56

[29] Geens T, Aerts D, Berthot C, Bourguignon JP, Goeyens L, Lecomte P, Maghuin-Rogister G, Pironnet AM, Pussemier L, Scippo ML, Van Loco J, Covaci A. A review of dietary and non-dietary exposure to bisphenol-A. Food and Chemical Toxicology. 2012;**50**(10):3725-3740

[30] Goodson A, Robin H, Summerfield W, Cooper I. Migration of 41bisphenol A from can coatings–Effects of damage, storage conditions and heating. Food Additives and Contaminants. 2004;**21**(10):1015-1026

[31] Calafat AM, Weuve J, Ye X, Jia LT, Hu H, Ringer S, Huttner K, Hauser R. Exposure to bisphenol A and other phenols in neonatal intensive care unit premature infants. Environmental Health Perspectives. 2009;**117**(4):639-644

[32] Calafat AM, Ye X, Wong LY, Reidy JA, Needham LL. Exposure of the U.S. population to bisphenol A and 4-tertiary-octylphenol: 2003–2004. Environmental Health Perspectives. 2008;**116**(1):39-44

[33] Braun JM, Kalkbrenner AE, Calafat AM, Bernert JT, Ye X, Silva MJ, Barr DB, Sathyanarayana S, Lanphear BP. Variability and predictors of urinary bisphenol A concentrations during pregnancy. Environmental Health Perspectives. 2011;**119**(1):131-137

[34] Vandenberg LN, Hauser R, Marcus M, Olea N, Welshons WV. Human exposure to bisphenol A (BPA). Reproductive Toxicology. 2007;**24**(2):139-177. Review

[35] Engel LS, Buckley JP, Yang G, Liao LM, Satagopan J, Calafat AM, Matthews CE, Cai Q, Ji BT, Cai H, Engel SM, Wolff MS, Rothman N, Zheng W, Xiang YB, Shu XO, Gao YT, Chow WH. Predictors and variability of repeat measurements of urinary phenols and parabens in a cohort of Shanghai women and men. Environmental Health Perspectives. 2014;**122**(7):733-740

[36] Sun Y, Irie M, Kishikawa N, Wada M, Kuroda N, Nakashima K. Determination of bisphenol A in human breast milk by HPLC with column-switching and fluorescence detection. Biomedical Chromatography. 2004;**18**(8):501-507

[37] Ye X, Kuklenyik Z, Needham LL, Calafat AM. Measuring environmental phenols and chlorinated organic chemicals in breast milk using automated on-line column-switching-high performance liquid chromatography-isotope dilution tandem mass spectrometry. Journal of Chromatography B, Analytical Technologies in the Biomedical and Life Sciences. 2006;**831**(1–2):110-115

[38] Ikezuki Y, Tsutsumi O, Takai Y, Kamei Y, Taketani Y. Determination of bisphenol A concentrations in human biological fluids reveals significant early prenatal exposure. Human Reproduction. 2002;**17**(11):2839-2841

[39] Schönfelder G, Wittfoht W, Hopp H, Talsness CE, Paul M, Chahoud I. Parent bisphenol A accumulation in the human maternal-fetal-placental unit. Environmental Health Perspectives. 2002;**110**(111):A703-A707

[40] World Health Organization. BISPHENOL A (BPA)–Current state of knowledge and [Internet]. 27 November 2009. Available from: http://www.who.int/foodsafety/publications/fs_management/No_05_Bisphenol_A_Nov09_en.pdf

[41] European Food Safety Authority. Opinion of the Scientific Panel on Food Additives, Flavourings, Processing Aids and Materials in Contact with Food on a request from the Commission related to 2,2-bis(4-hydroxyphenyl)propane [Internet]. 29 November 2006. Available from: http://www.efsa.europa.eu/sites/default/files/scientific_output/files/main_documents/428.pdf

[42] Snyder RW, Maness SC, Gaido KW, Welsch F, Sumner SC, Fennell TR. Metabolism and disposition of bisphenol A in female rats. Toxicology and Applied Pharmacology. 2000;**168**:225-234

[43] Matthews JB, Twomey K, Zacharewski TR. *In vitro* and *in vivo* interactions of bisphenol A and its metabolite, bisphenol A glucuronide, with estrogen receptors alpha and beta. Chemical Research in Toxicology. 2001;**14**(2):149-157

[44] Pottenger LH, Domoradzki JY, Markham DA, Hansen SC, Cagen SZ, Waechter JM Jr. The relative bioavailability and metabolism of bisphenol A in rats is dependent upon the route of administration. Toxicological Sciences. 2000;**54**(1):3-18

[45] Stahlhut RW, Welshons WV, Swan SH. Bisphenol A data in NHANES suggest longer than expected half-life, substantial nonfood exposure, or both. Environmental Health Perspectives. 2009;**117**(5):784-789

[46] Völkel W, Colnot T, Csanády GA, Filser JG, Dekant W. Metabolism and kinetics of bisphenol A in humans at low doses following oral administration. Chemical Research in Toxicology. 2002;**15**(10):1281-1287

[47] Thayer KA, Doerge DR, Hunt D, Schurman SH, Twaddle NC, Churchwell MI, Garantziotis S, Kissling GE, Easterling MR, Bucher JR, Birnbaum LS. Pharmacokinetics of bisphenol A in humans following a single oral administration. Environmental International. 2015;**83**:107-115

[48] Pritchett JJ, Kuester RK, Sipes IG. Metabolism of bisphenol A in primary cultured hepatocytes from mice, rats, and humans. Drug Metabolism and Disposition. 2002;**30**(11):1180-1185

[49] Teeguarden JG, Twaddle NC, Churchwell MI, Yang X, Fisher JW, Seryak LM, Doerge DR. 24-hour human urine and serum profiles of bisphenol A: Evidence against sublingual absorption following ingestion in soup. Toxicology and Applied Pharmacology. 2015;**288**(2):131-142

[50] Domoradzki JY, Thornton CM, Pottenger LH, Hansen SC, Card TL, Markham DA, Dryzga MD, Shiotsuka RN, Waechter JM Jr. Age and dose dependency of the pharmacokinetics and metabolism of bisphenol A in neonatal Sprague–Dawley rats following oral administration. Toxicological Sciences. 2004;**77**(2):230-242

[51] Matsumoto J, Yokota H, Yuasa A. Developmental increases in rat hepatic microsomal UDP-glucuronosyltransferase activities toward xenoestrogens and decreases during pregnancy. Environmental Health Perspectives. 2002;**110**(2):193-196

[52] Gould JC, Leonard LS, Maness SC, Wagner BL, Conner K, Zacharewski T, Safe S, McDonnell DP, Gaido KW. Bisphenol A interacts with the estrogen receptor alpha in a distinct manner from estradiol. Molecular and Cellular Endocrinology. 1998;**142**(1–2):203-214

[53] Kuiper GG, Lemmen JG, Carlsson B, Corton JC, Safe SH, van der Saag PT, van der Burg B, Gustafsson JA. Interaction of estrogenic chemicals and phytoestrogens with estrogen receptor beta. Endocrinology. 1998;**139**(10):4252-4263

[54] Welshons WV, Thayer KA, Judy BM, Taylor JA, Curran EM, vom Saal FS. Large effects from small exposures. I. Mechanisms for endocrine-disrupting chemicals with estrogenic activity. Environmental Health Perspectives. 2003;**111**(8):994-1006. Review

[55] Ziv-Gal A, Craig ZR, Wang W, Flaws JA. Bisphenol A inhibits cultured mouse ovarian follicle growth partially via the aryl hydrocarbon receptor signaling pathway. Reproductive Toxicology. 2013;**42**:58-67

[56] Melnick R, Lucier G, Wolfe M, Hall R, Stancel G, Prins G, Gallo M, Reuhl K, Ho SM, Brown T, Moore J, Leakey J, Haseman J, Kohn M. Summary of the National Toxicology Program's report of the endocrine disruptors low-dose peer review. Environmental Health Perspectives. 2002;**110**(4):427-431

[57] Hoover RN, Hyer M, Pfeiffer RM, Adam E, Bond B, Cheville AL, Colton T, Hartge P, Hatch EE, Herbst AL, Karlan BY, Kaufman R, Noller KL, Palmer JR, Robboy SJ, Saal RC, Strohsnitter W, Titus-Ernstoff L, Troisi R. Adverse health outcomes in women exposed in utero to diethylstilbestrol. The New England Journal of Medicine. 2011;**265**(14):1304-1314

[58] Markey CM, Coombs MA, Sonnenschein C, Soto AM. Mammalian development in a changing environment: Exposure to endocrine disruptors reveals the developmental plasticity of steroid-hormone target organs. Evolution & Development. 2003;**5**(1):67-75

[59] Wetherill YB, Akingbemi BT, Kanno J, McLachlan JA, Nadal A, Sonnenschein C, Watson CS, Zoeller RT, Belcher SM. *In vitro* molecular mechanisms of bisphenol A action. Reproductive Toxicology. 2007;**24**(2):178-198. Review

[60] Wang IJ, Chen CY, Bornehag CG. Bisphenol A exposure may increase the risk of development of atopic disorders in children. International Journal of Hygiene and Environmental Health. 2016;**219**(3):311-316

[61] Miao M, Yuan W, Yang F, Liang H, Zhou Z, Li R, Gao E, Li DK. Associations between bisphenol A exposure and reproductive hormones among female workers. International Journal of Environmental Research and Public Health. 2015;**12**(10):13240-13250

[62] Li DK, Miao M, Zhou Z, Wu C, Shi H, Liu X, Wang S, Yuan W. Urine bisphenol-A level in relation to obesity and overweight in school-age children. PLoS One. 2013;**8**(6):e65399

[63] Vandenberg LN, Colborn T, Hayes TB, Heindel JJ, Jacobs DR Jr, Lee DH, Shioda T, Soto AM, vom Saal FS, Welshons WV, Zoeller RT, Myers JP. Hormones and endocrine-disrupting chemicals: Low-dose effects and nonmonotonic dose responses. Endocr Rev. 2012 Jun;**33**(3):378-455. Review

[64] Richter CA, Birnbaum LS, Farabollini F, Newbold RR, Rubin BS, Talsness CE, Vandenbergh JG, Walser-Kuntz DR, vom Saal FS. *In vivo* effects of bisphenol A in laboratory rodent studies. Reproductive Toxicology. 2007;**24**(2):199-224. Review

[65] Ryan KK, Haller AM, Sorrell JE, Woods SC, Jandacek RJ, Seeley RJ. Perinatal exposure to bisphenol-A and the development of metabolic syndrome in CD-1 mice. Endocrinology. 2010;**151**(6):2603-2612

[66] Meeker JD, Ehrlich S, Toth TL, Wright DL, Calafat AM, Trisini AT, Ye X, Hauser R. Semen quality and sperm DNA damage in relation to urinary bisphenol A among men from an infertility clinic. Reproductive Toxicology. 2010;**30**(4):532-539

[67] Melzer D, Harries L, Cipelli R, Henley W, Money C, McCormack P, Young A, Guralnik J, Ferrucci L, Bandinelli S, Corsi AM, Galloway T. Bisphenol A exposure is associated with *in vivo* estrogenic gene expression in adults. Environmental Health Perspectives. 2011;**119**(2):1788-1793

[68] Patisaul HB, Adewale HB. Long-term effects of environmental endocrine disruptors on reproductive physiology and behavior. Frontiers in Behavioral Neuroscience. 2009;**3**:10

[69] Barker DJ. The origins of the developmental origins theory. Journal of Internal Medicine. 2007;**261**(5):412-417. Review

[70] Doherty LF, Bromer JG, Zhou Y, Aldad TS, Taylor HS. In utero exposure to diethylstilbestrol (DES) or bisphenol-A (BPA) increases EZH2 expression in the mammary gland: an epigenetic mechanism linking endocrine disruptors to breast cancer. Hormones & Cancer. 2010;**1**(3):146-155

[71] Inagaki T, Smith N, Lee EK, Ramakrishnan S. Low dose exposure to bisphenol A alters development of gonadotropin-releasing hormone 3 neurons and larval locomotor behavior in Japanese Medaka. Neurotoxicology. 2016;**52**:188-197

[72] Roy A, Bauer SM, Lawrence BP. Developmental exposure to bisphenol A modulates innate but not adaptive immune responses to influenza A virus infection. PLoS One. 2012;**7**(6):e38448

[73] Kjaerstad MB, Taxvig C, Andersen HR, Nellemann C. Mixture effects of endocrine disrupting compounds *in vitro*. International Journal of Andrology. 2010;**33**(2):425-433

[74] Metzdorff SB, Dalgaard M, Christiansen S, Axelstad M, Hass U, Kiersgaard MK, Scholze M, Kortenkamp A, Vinggaard AM. Dysgenesis and histological changes of genitals and perturbations of gene expression in male rats after in utero exposure to antiandrogen mixtures. Toxicological Sciences. 2007;**98**(1):87-98

[75] Berenbaum MC. The expected effect of a combination of agents: the general solution. Journal of Theoretical Biology. 1985;**114**(3):413-431

[76] Yang YJ, Hong YP. Combination effects of bisphenol A and isobutylparaben on the green macroalga Ulva pertusa. Toxicology and Environmental Health Sciences. 2012;**4**(1):37-41

[77] Kim YR, Jung EM, Choi KC, Jeung EB. Synergistic effects of octylphenol and isobutyl paraben on the expression of calbindin-D_9k in GH3 rat pituitary cells. International Journal of Molecular Medicine. 2012;**29**(2):294-302

[78] Manikkam M, Tracey R, Guerrero-Bosagna C, Skinner MK. Plastics derived endocrine disruptors (BPA, DEHP and DBP) induce epigenetic transgenerational inheritance of obesity, reproductive disease and sperm epimutations. PLoS One. 2013;**8**(1):e55387

Chapter 7

The Toxic Effects BPA on Fetuses, Infants, and Children

Mujtaba Ellahi and Mamoon ur Rashid

Additional information is available at the end of the chapter

http://dx.doi.org/10.5772/intechopen.68896

Abstract

Bisphenol A (BPA) is an organic synthetic compound with the chemical formula $(CH_3)_2C(C_6H_4OH)_2$ belonging to the group of diphenylmethane derivatives and bisphenols, with two hydroxyphenyl groups. BPA is the common name for 2,2-(4,4′-dihydroxydiphenyl) propane, IUPAC name 4,4′-(propane-2,2-diyl) diphenol, alternative name *p,p′*-isopropylidenebisphenol, with two phenol moieties. Its important properties include low vapor pressure, moderate water solubility, and low volatility. It is a colorless solid that is soluble in organic solvents, but poorly soluble in water. BPA is a plastic component produced in large quantities for use chiefly in the production of polycarbonate plastics and epoxy resins. BPA epoxy has a good, broad range of chemical resistance, good physical properties, and is cured using a wide variety of curing agents at ambient temperatures. The present chapter focuses on different toxic effects and the influence of BPA on different stages of human life in fetuses, infants, and children. The chapter also concentrates on how to handle BPA, its treatment, and preventive measures against BPA exposure.

Keywords: toxic effects, fetuses, infants, children

1. Introduction

Bisphenol A (BPA) is a monomer used in polymer plastic material and is used comprehensively in the manufacture of polycarbonate plastics and epoxy resins. More specifically, food packaging bottles are mostly made of polycarbonate plastics, though the resins are usually used for polishing and coating of metal products including food cans, bottle tops, and water supply pipes. BPA is also used in the manufacture of polyacrylate resins, polysulfone resins, polyester resins, flame retardants, and in the recycling of thermal paper. Dental sealants and polymeric tooth coatings also contain BPA [1]. We are exposed BPA toxicity directly through its products or indirectly throughout contaminated surroundings. Diet (food and beverages) is the chief source of human exposure to BPA, though air, dust, and water (including skin contact during bathing and swimming) are other possible sources of exposure. It was also noted that thermal printing paper (cashier's receipts) causes contamination of human skin with BPA. However, individuals employed in the BPA industry can be exposed directly to BPA at the workplace [2]. Leaching of BPA from polycarbonate products depends on the contact time, temperature, and type of food. It takes place either through the hydrolysis (breaking H^+ and OH^- ions) of residual BPA in polycarbonate products after they are developed or through diffusion in the case of dry food products. The presence of food simulators as 50% ethanol (C_2H_5OH) or 3% acetic acid (CH_3COOH) also causes enhanced leaching of BPA in food [3]. BPA may be the bona fide endocrine disruptor that unfavorably affects metabolic homeostasis. Endocrine disrupting chemicals (EDCs) have the capability of hindering normal endocrine systems. The two EDCs, BPA and triclosan (TCS), are mass produced and widespread.

2. Discussion

2.1. Scientific concerns over BPA

There are numerous scientific concerns regarding BPA toxicity in humans, particularly in fetuses, infants, and children. The data regarding BPA distribution in the environment and potential for human exposure raise public awareness and concern about BPA. The production capacity and solid waste management of BPA products and extent of the database on biochemical properties of BPA in fetuses, infants, and children are yet to be compiled from various other databases and literature sources. The detailed mechanism of BPA enzymatic, androgenic, and neurological alterations, hematological effects and histological toxicity in fetuses, infants, and children is not convincing because of inadequate studies or traditional strategies carried out on individuals exposed to BPA [4]. It is very important to plan how to handle, dispose of, and implement protective measures against BPA exposure. There is restrained doubt regarding BPA toxicity because industries are trying their best to defend and demonstrate that BPA is not that dangerous, and some researchers have proved that BPA is less toxic than was originally thought. However, it will take a long time to fully ban BPA production and usage or introduce an alternative to the plastics industries. Because of diverse opinion, some standard procedures will have to be established for the safe removal of BPA from the environment rather than imposing a complete ban [5]. In developed countries, BPA is ever present and has been detected in nearly all human serum samples [6]. In different parts of the world, BPA has been detected in human saliva, serum, urine, amniotic fluid, umbilical cord blood, and placental tissues. BPA has also been detected in human nails, hair, the dermis, breast, and in subcutaneous and visceral adipose tissue. Most of the scientific studies (over 130) have reported seven harmful effects of BPA, including breast cancer, early puberty, heart diseases, infertility in males and females, as a catalyst for multiple negative brain variations, and obesity [7].

2.2. BPA toxicity mechanism

There are several mechanisms by which BPA affects human health. In receptor-mediated mechanisms, endocrine disturbance takes place by BPA capability to act directly as ligands (agonist) for steroid hormone nuclear receptors (NRs), in particular estrogen, androgen, and thyroid hormone receptors. In nonreceptor-mediated mechanisms, the cytochrome P450 enzymes responsible for the highly specific reactions in the steroid biosynthesis pathway are some of the molecular targets of interest, given their vital role in the formation of different highly effective endogenous steroid hormones. BPA interferes with steroidogenic enzymes and hormone transport in the form of BPA as both an androgen receptor antagonist and polychlorinated biphenyls (PCBs). BPA has been listed among the EDCs and is known for its metabolic homeostasis disturbance to block endogenous hormonal activity. Endocrine disturbance is connected with the risk of developmental problems, cancer, diabetes, obesity, and the metabolic syndrome. Also endocrine disruptors can lead to sexual sterility and interfere with fertility. BPA also causes both disruption in the function and structure of the brain and irregularities in the flow of hormones from the brain, which controls and regulates life processes. However, the probable developmental or reproductive hazard is not convincing because of incomplete data [8, 9]. Nonetheless, BPA exerts unexpected exposure outcome relationships, because low doses frequently exert stronger toxicity than higher doses because of its estrogenic and other biotic features [10]; BPA ≤5 mg/kg body weight/day during the critical phases of growth might affect healthiness later in life [11]. From a toxic dynamics point of view, it was investigated that BPA half-life was relatively short and thus it was suggested to be a fairly less cumulative chemical. Following oral exposure, most of the BPA dose absorbed was glucuronidated in rat liver and intestines, and because of the short half-life of BPA, this biochemical was excreted quickly from the body by making BPA sulfates $(SO_4)^{-2}$ or BPA glucuronidates during the metabolic progression [6, 12].

2.3. BPA toxicity and different human life stages

Numerous effects of BPA, both in humans and in animals, have been extensively studied and the target anatomy (organs) has been identified in repeat-dose bodily screening of liver, intestines, and kidneys. The effect of BPA in producing hormonal disturbance and potential associated problems of neurological, epidemiological, physical, and behavioral development were analyzed in fetuses, infants, and children [13]. A number of properties of BPA in animals have been widely examined, and objective structures recognized in repeat-dose animal studies contain intestines, liver, and kidneys. Conversely, the main concern has been those effects related to the hormonal movement of BPA and possibly the connected effects on physical, neurological, and developmental growth.

The discussion regarding the human health effects of BPA contact is restricted by a deficiency of epidemiological statistics. At present, there is not enough arithmetical influence to calculate dose effects or decide all the health penalties of exposure to BPA in humans [13]. Frequently, exposure is measured as being steady over time, with the total dose predictable by increasing exposure; it is computed as the product of attentiveness. Nevertheless, if contact is not steady over time, the similar total increasing contact delivered in various forms may create several organic effects.

2.4. BPA toxicity in the fetal stage

In the fetal stage of the human lifecycle, there is serious concern that BPA may enter the human placenta, exposing the fetus, and the consequences of BPA toxicity of "estrogenic chemicals" is more severe in the developing fetus as compared to the adult organism.

Estrogenic chemicals have the potential to restrict the steroid-dependent body of the neuroendocrine and neurochemical systems. It is in the evolving brain that variations in the estrogenic environment affect several features of cellular reproduction, together with neuritis flexibility and branching, synaptic development, expression of neurotransmitters, cell survival, and death. Most of the human data, even though inconsistent, show that prenatal BPA exposure could seriously affect child sex-dependent behaviors [14, 15]. Many studies have recommended that BPA exposure is connected with female infertility. However, the relationship between TCS exposure and female infertility remains unidentified. Mice have been used as an animal model to study the relationship between exposure to these two chemicals and infertility [16, 17]. Slight changes in estrogen levels can lead to implantation collapse in humans and mice [17–19]. BPA and TCS have estrogenic movement in vitro and in vivo [20, 21]. BPA binds to both Estrogen receptor (ER) [Estrogen receptors (ERs) are a group of proteins found inside and on cells. They are receptors that are activated by the hormoneestrogen (17β-estradiol)] [22, 23]. Equally, BPA and TCS have many biological effects mediated via estrogen receptors [24, 25]. Therefore BPA and TCS may source implantation collapse because of their capability to mimic estrogen in humans [26, 27]. In humans, from oocyte maturation to implantation, the organic features of the oocyte and the embryo vary noticeably. The levels of sex hormones, such as estrogen, progesterone, and androgen, and their receptors also alter significantly. Consequently, consideration of the female reproductive system to BPA and TCS may vary depending on the time of contact. It has been reported that in mice, preimplantation exposure to a similar quantity of BPA or TCS on gestational days 2 and 3 is stronger to induce embryo implantation failure than exposure on gestational days 0 and 1 [28–33]. Thus in mice, gestational days 2 and 3 may be a susceptible window for BPA and TCS. Exposure to these two endocrine disruptors during a susceptible window might lead to implantation stoppage. On the other hand, in humans, the susceptible window for these EDCs still needs further investigation.

In the health center, fertilization can only be established in one way. There is a way to identify embryo implantation collapse in the health center except for patients undergoing in vitro fertilization. The most common way to analyze pregnancy is by testing human chorionic gonadotrophin (HCG) in urine samples. However, HCG is secreted by the syncytiotrophoblast and is noticeable in maternal blood 2 days after implantation of the embryo [30]. Therefore it is possible that many women did not know that they had a fertilized embryo, which subsequently failed to implant into their endometrium because no HCG was secreted. Yet, if there was vaginal bleeding and a gynecologist was consulted, it will only be seen as ovulation bleeding, which is ordinary common occurance in the health center. More often than not, when a woman wants to know if she is pregnant she will perform a urine pregnancy test. On the other hand, a measurable level of HCG in the urine requires the embryo to survive for at least a week after implantation. Since BPA and TCS can be engrossed and excreted rapidly and do not build up in the human body [34–39], a change in habits such as stopping the use of TCS-containing toothpaste or plastic food containers can result in a variation in the levels of these two chemicals in the human body. This means that if these habits cease for the duration of the susceptible timeframe—for example, the woman no longer uses TCS-containing toothpaste or is using novel TCS-free toothpaste, or no longer uses plastic containers that leak BPA—it could result in a comparatively low level of TCS and BPA in her body and pregnancy could ensue. In the health center, a woman cannot be assessed as infertile unless she has attempted defenseless coitus for at least 1 year without becoming pregnant. This means that maybe BPA and TCS have caused more miscarriages than have been realized. Furthermore, the most susceptible time for BPA and TCS to pursue implantation remains indefinite. Even though the preimplantation period might be a susceptible timeframe for BPA and TCS contact, it might not be the most responsive and significant.

2.5. BPA toxicity in children

Studies have concluded that BPA affects children's health, and shown links between parental urinary BPA concretion and depressive, anxiety, and hyperactive behaviors in children (2–3 years); however, the results were more pronounced for girls than for boys Braun et al. [40]. On the contrary, Perera et al. declared reduced nervous, depressed, and hostile behaviors in girls (3–5 years), but violent and emotionally sensitive behaviors in boys (3–5 years) with prenatal exposure to BPA (measured in pregnant mothers' urine). The BPA variant noncoplanar PCB affects dopamine, serotonin, and acetylcholine, while coplanar PCB affects thyroid hormone and glucocorticoids, and the consequence of these effects are more severe in fetuses, infants, and children [41].

2.6. BPA toxicity in adults

Recently, a number of rare studies have detected the relations between BPA exposure and growth and the reproduction syndrome in humans. Investigations on humans have shown the connection between urine, feces, or blood concentrations of BPA (total/free) and a diversity of health measures including mutation in fetuses, miscarriage, obesity and fertility in women, effect on the uterus lining ("endometrium"), certain hormones that support the control of deoxyribonucleic acid (DNA) damage to reproduction markers, length of gestation, polycystic ovary syndrome, and birth outcomes [42]. Only a small number of studies have summarized the relations between BPA exposure and disorders of reproduction or developmental effects in humans. Studies on humans have looked at the association between urine or blood concentrations of total or free BPA and a variety of incorporated health measures similar to:

i. hormones that help to control reproduction markers of DNA injury;

ii. miscarriage or mishandling;

iii. infection in fetuses;

iv. fertility and obesity in women;

v. properties of the tissue of the uterus ("endometrium");

vi. polycystic ovary syndrome, delivery outcomes, and extent of development.

3. Treatment

After ingestion, a metabolic development is excreted by enzymes first and foremost in the liver where the majority of BPA is bound quickly to glucuronic acid to create BPA glucuronide. Because BPA is rapidly solvable in water (H_2O), it is better to get rid of BPA in the urine, which also reduces its capability to interrelate with organic processes in the body. When rats were exposed to BPA in their food with probiotics, their BPA blood concentrations dropped considerably and were defecated 2.7 times more firmly than the nonsupplemented organized collection. In other words, probiotics decreased intestinal amalgamation by boosting BPA secretion and may also repress BPA's unfavorable effects on human health.

A simple and cheap way to increase the probiotic content of energy expenditure is to take supplements such as IVL's Flora Life, the first and only acid-proof, suspended-discharge probiotic accessible on the open market nowadays. One or two capsules a day are all it takes to introduce the digestive zone to 22 billion acceptable bacteria.

Once in the intestine, these suspended free capsules carry the live probiotics essential to vigorously stabilize the gut bacteria and help free the body of poisonous chemicals, along with other numerous health advantages.

4. Preventive measures

4.1. For infants

As investigations continue, concerned parents can take the following preventive actions to decrease infant contact with BPA:

i. avoid pristine plastic infant bottles or containers with the recycling no. 7 and the letters "PC" imprinted on them; a number of these contain BPA;

ii. use licensed or recognized BPA-free synthetic bottles;

iii. use bottles made of opaque plastic, i.e., those made of polyethylene terephthalate (PET) or polypropylene; these do not contain BPA and have the recycling numbers 2 or 5 on them;

Because heat may be a source of BPA leakage from plastic, the following should be noted:

iv. do not heat or boil polycarbonate bottles;

v. do not wash polycarbonate flasks in the dishwasher;

vi. glass flasks can be an option; however, if the flask is dropped it may break;

vii. breastfeeding is an additional method to decrease probable BPA exposure;

viii. risks connected with giving infants unsuitable (home-based condensed milk) formulas or substitute (soy, goat, or sheep) milk is better than the possible effects of BPA.

4.2. For children and adults

i. before using dental sealants, ask the dentist if the ingredients in the products they use contain BPA;

ii. use glass, stainless steel, paper, cloth, or clay containers for food and beverages;

iii. be aware that "microwave safe and sound" only means that the container or cling wrap will not distort; it has nothing to do with protection;

iv. avoid contact with oily or sour foods and synthetic organisms;

v. recycle any damaged or dented synthetic items;

vi. use ceramic or stainless steel containers;

vii. check the type of plastic a food processor is made from; substitute synthetic coffee filters with clay or metal ones.

5. Conclusion

The current research establishes no proof that BPA is a growing neurotoxicant. Concern is growing regarding the use of BPA products used all over the world. However, it will take a long time to completely prohibit the use of BPA in the synthetics industries, because they are striving to justify and establish that BPA is not that harmful, as shown by a variety of studies and investigations throughout the world. The synthetics industries have a dissimilar viewpoint and as a result much effort is necessary to change methods for the secure elimination of BPA from the environment rather than aiming at banning its use altogether. Interior coverage to free BPA accessible for organic activity within the body is thus probable to be extremely low down. Newborns are susceptible to upper interior BPA standards because of undeveloped glucuronidation movement. Numerous studies in adult women report a relation between BPA exposure and effects on the reproductive system, e.g., recurrent miscarriages, endometrial hyperplasia, and polycystic ovary syndrome. BPA exposure can disturb pubertal timing and cause irregular ovulatory cycles in rodents and these defects result from the abnormal organization of the hypothalamic– pituitary–gonadal axis, the central neuroendocrine corridor that regulates the reproductive process. BPA has also been found to induce apoptosis. As research continues, anxious parents can take preventive measures to diminish infants' exposure to BPA: regular breastfeeding, avoiding the use of dishwashers, and heating or boiling polycarbonate bottles or using BPA-free plastic bottles made of polyethylene, polypropylene, or glass. Children and adults should be encouraged to use glass, paper, cloth, stainless steel, or ceramic packaging/bottles for food and beverages.

There is considerable proof that endocrine disruptors are linked to cancer, childhood development, diabetes, and probably also obesity and metabolic conditions. In addition, it seems extremely possible that endocrine disruptors can add to sterility and associate with fertility. Scientific choices regarding health perils are usually based on what is recognized as the "proof mass." Proof from the incomplete number of research studies in humans exposed to BPA is insufficient to reach conclusions concerning probable developmental or reproductive risk.

These discrepancies deserve additional investigations for enhanced acceptance of toxic kinetics, class, and interindividual changes, likely for additional sources of contact with BPA and possible confounders impacting on the consequences. An extensive research study can even cover the method for the growth of probiotics, i.e., live microbes that, when administered in sufficient amounts, present a health advantage to the host. Conceivably, these probiotics could be used for the secure elimination of accumulated BPA from live systems.

Abbreviations

BPA	Bisphenol A
EDCs	Endocrine disrupting chemicals
TCS	Triclosan
NRs	Nuclear receptors
PCBs	Polychlorinated biphenyls
HCG	Human chorionic gonadotrophin
DNA	Deoxyribonucleic acid
PET	Polyethylene terephthalate

Author details

Mujtaba Ellahi* and Mamoon ur Rashid

*Address all correspondence to: mujtaba.ellahi@gmail.com

Department of Chemistry, Faculty of Arts and Basic Sciences, Balochistan University of Information Technology, Engineering and Management Sciences (BUITEMS), Quetta, Pakistan

References

[1] Beronius A., Hanberg A. (2011). Sources of exposure to bisphenol A. IMM report 2/2011. Institute of Environmental Medicine, Karolinska Institute: ki.se/IMM.

[2] Daniel Z., Carine J., Hélène D., Sandrine B., Elisabeth P. (2011). Viable skin efficiently absorbs and metabolizes bisphenol A. Chemosphere, 82(3), 424–430.

[3] Le H. H., Carlson E. M., Chua J. P., Belcher S. M. (2008). Bisphenol A is released from polycarbonate drinking bottles and mimics the neurotoxic actions of estrogen in developing cerebellar neurons. Toxicology Letters, 176(2), 149–156.

[4] Kanga J.-H., Kondob F., Katayamaa Y. (2006). Human exposure to bisphenol A. Toxicology journal, 226(2–3), 79–89.

[5] Sachs N. M. (2012). Protecting the public from BPA: an action plan for federal agencies. University of Richmond, Law of faculty publication; Center for Progressive Reform white paper #1202.

[6] Domoradzki J. Y., Pottenger L. H., Thornton C. M., Hansen S. C., Card T.L., Markham D. A., Dryzga, M. D., Shiotsuka R. N., Waechter J. M. (2003). Metabolism and Pharmacokinetics of Bisphenol A (BPA) and the Embryo-Fetal Distribution of BPA and BPA-Monoglucuronide in CD Sprague-Dawley Rats at Three Gestational Stages. Journal of Toxicological Sciences, 76, 21–34.

[7] Pritchett J. J., Kuester R. K., Sipes I. G. (2002). Metabolism of bisphenol a in primary cultured hepatocytes from mice, rats, and humans. Drug Metabolism and Disposition, 30, 1180–1185.

[8] Doherty L. F., Bromer J. G., Zhou Y., Aldad T. S., Taylor H. S. (2010). In utero exposure to diethylstilbestrol (DES) or bisphenol-A (BPA) increases EZH2 expression in the mammary gland: an epigenetic mechanism linking endocrine disruptors to breast cancer. Hormones and Cancer, 1(3), 146–155.

[9] Vom Saal F. S., Akingbemi B. T., Belcher S. M., Birnbaum L. S., Crain D. A., Eriksen M., Farabollini F. et al. (2007). Chapel Hill bisphenol A expert panel consensus statement: integration of mechanisms, effects in animals and potential to impact human health at current levels of exposure. Reproductive Toxicology (Elmsford, NY), 24(2), 131.

[10] Bottini, A. T., Bowen, J. R., Gilchrist, D. G. (1981). Phytotoxins. II. Characterization of a phytotoxic fraction from Alternaria alternata f. sp. lycopersici. Tetrahedron Letters, 22, 2723–2726.

[11] Shin S. D., Lee B. S., Lee B. M., Han K. C., Kim S. Y., Kwack H. S., Park S. J. (2001). Bioavailability and mammary excretion of bisphenol a in sprague-dawley rats. Journal of Toxicology and Environmental Health A, 64, 417–426.

[12] Pottenger L. H., Domoradzki J. Y., Markham D. A., Hansen S. C., Cagen S. Z., Waechter J. M. (2000). The Relative Bioavailability and Metabolism of Bisphenol A in Rats Is Dependent upon the Route of Administration. Journal of Toxicological Sciences, 54, 3–18.

[13] Martan R., Jiménez E., Olivares M., Marin M.L., Fernande L., Xaus J., Rodriguez J.M. (2006). Lactobacillus salivarius CECT 5713, a potential probiotic strain isolated from infant feces and breast milk of a mother–child pair. International Journal of Food Microbiology, 112 112e–135e.

[14] Lang I. A., Galloway T. S., et al. (2008). Association of Urinary Bisphenol A Concentration With Medical Disorders and Laboratory Abnormalities in Adults. JAMA 300, 1303–1310.

[15] Yolton K., Xu Y., Strauss D., Altaye M., Calafat A.M., Khoury J. (2008). Prenatal exposure to bisphenol A and phthalates and infant neurobehavior. Neurotoxicology and Teratology 33, 558–566.

[16] Xiao S., Diao H., Smith M. A., Song X., Ye X. (2011). Preimplantation exposure to bisphenol A (BPA) affects embryo transport, preimplantation embryo development, and uterine receptivity in mice. Reproductive Toxicology, 32(4), 434–441.

[17] Crawford B. R., de-Catanzaro D. (2012). Disruption of blastocyst implantation by triclosan in mice: impacts of repeated and acute doses and combination with bisphenol-A. Reproductive Toxicology, 34(4), 607–613.

[18] Berger R. G., Foster W. G., deCatanzaro D. (2010). Bisphenol-A exposure during the period of blastocyst implantation alters uterine morphology and perturbs measures of estrogen and progesterone receptor expression in mice. Reproductive Toxicology, 30(03), 393–400.

[19] Berger R. G., Hancock T., deCatanzaro D. (2007). Influence of oral and subcutaneous bisphenol-A on intrauterine implantation of fertilized ova in inseminated female mice. Reproductive Toxicology, 23(02), 138–144.

[20] Berger R. G., Shaw J., deCatanzaro D. (2008). Impact of acute bisphenol-A exposure upon intrauterine implantation of fertilized ova and urinary levels of progesterone and 17 –estradiol. Reproductive Toxicology, 26(02), 94–99.

[21] Ma W.-G., Song H., Das S. K., Paria B. C., Dey S. K. (2003). Estrogen is a critical determinant that specifies the duration of the window of uterine receptivity for implantation. Proceedings of the National Academy of Sciences of the United States of America, 100(05), 2963–2968.

[22] Gidley-Baird A. A., O'Neill C., Sinosich M. J., Porter R. N., Pike I. L., Saunders D. M. (1986). Failure of implantation in human in vitro fertilization and embryo transfer patients: the effects of altered progesterone/estrogen ratios in humans and mice. Fertility and Sterility, 45(01), 69–74.

[23] Jung E.-M., An B.-S., Choi K.-C., Jeung E.-B. (2012). Potential estrogenic activity of triclosan in the uterus of immature rats and rat pituitary GH3 cells. Toxicology Letters, 208(02), 142–148.

[24] Welshons W.V., Nagel S. C., vom Saal F. (2006). Large effects from small exposures. III. Endocrine mechanisms mediating effects of bisphenol A at levels of human exposure. Endocrinology, 147(06), supplement, S56–S69.

[25] Gould J. C., Leonard L., Maness S. et al. (1998). Bisphenol A interacts with the estrogen receptor alpha in a distinct manner from estradiol. Molecular and Cellular Endocrinology, 142(12), 203–214.

[26] Kuiper G. G. J. M., Lemmen J. G., Carlsson B. et al. (1998). Interaction of estrogenic chemicals and phytoestrogens with estrogen receptor. Endocrinology, 139(10), 4252–4263.

[27] Pennie W. D., Aldridge T. C., Brooks A. N. (1998). Differential activation by xenoestrogens of ER and ER when linked to different response elements. Journal of Endocrinology, 158(03), R11–R14.

[28] Lee H.-R., Hwang K.-A., Nam K.-H., Kim H.-C., Choi K.-C. (2014). Progression of breast cancer cells was enhanced by endocrine-disrupting chemicals, triclosan and octylphenol, via an estrogen receptor-dependent signaling pathway in cellular and mouse xenograft models. Chemical Research in Toxicology, 27(05), 834–842.

[29] Miyagawa S., Lange A., Hirakawa I. et al. (2014). Differing species responsiveness of estrogenic contaminants in fish is conferred by the ligand binding domain of the estrogen receptor. Environmental Science and Technology, 48(09), 5254– 5263.

[30] Xu X., Lu Y., Zhang G. et al. (2014). Bisphenol A promotes dendritic morphogenesis of hippocampal neurons through estrogen receptor-mediated ERK1/2 signal pathway. Chemosphere, 96, 129–137.

[31] Kim S. M., Jung E. M., An B. S. et al. (2012). Additional effects of bisphenol A and paraben on the induction of calbindin-D9K and progesterone receptor via an estrogen receptor pathway in rat pituitary GH3 cells. Journal of Physiology and Pharmacology, 63(05), 445–455.

[32] Rivas A., Lacroix M., Olea-Serrano F., Laos I., Leclercq G., Olea N. (2002). Estrogenic effect of a series of bisphenol analogues on gene and protein expression in MCF-7 breast cancer cells. Journal of Steroid Biochemistry and Molecular Biology, 82(01), 45–53.

[33] Stoker T. E., Gibson E. K., Zorrilla L. M. (2010). Triclosan exposure modulates estrogen-dependent responses in the female Wistar rat. Toxicological Sciences, 117(01), 45–53.

[34] Ishibashi H., Matsumura N., Hirano M. et al. (2004). Effects of triclosan on the early life stages and reproduction of medaka Oryzias latipes and induction of hepatic vitellogenin. Aquatic Toxicology, 67(02), 167–179.

[35] Taylor J. A., vom Saal F. S., Welshonsetal W. V. (2011). Similarity of bisphenol A pharmacokinetics in rhesus monkeys and mice: relevance for human exposure. Environmental Health Perspectives, 119(04), 422–430.

[36] Rodricks J. V., Swenberg J. A., Borzelleca J. F., Maronpot R. R., Shipp A. M. (2010). Triclosan: a critical review of the experimental data and development of margins of safety for consumer products. Critical Reviews in Toxicology, 40(05), 422–484.

[37] Völkel W., Colnot T., Csanády G. A., Filser J. G., Dekant W. (2002). Metabolism and kinetics of bisphenol A in humans at low doses following oral administration. Chemical Research in Toxicology, 15(10), 1281–1287.

[38] Völkel W. V., Bittner N., Dekant W. (2005). Quantitation of bisphenol A and bisphenol A glucuronide in biological samples by high performance liquid chromatography-tandem mass spectrometry. Drug Metabolism and Disposition, 33(11), 1748–1757.

[39] Fournier T., Guibourdenche J., and Evain-Brion D (2015). Review: hCGs: different sources of production, different glycoforms and functions. Placenta, 36, S60–S65.

[40] Braun J. M., Kalkbrenner A. E., Calafat A. M., Bernert J. T., Ye X, et al. (2011). Variability and predictors of urinary bisphenol a concentrations during pregnancy. Environment Health Perspective, 119(01), 131–137.

[41] Braun J. M., Yolton K., Dietrich K. N., Hornung R., Ye X., Calafat A. M., Lanphear B. P. (2009). Prenatal Bisphenol A Exposure and Early Childhood Behavior. Environmental Health Perspective, 117, 1945–1952.

[42] Perera F., Tang D. L., Vishnevetsky J., Herbstman J., Calafat A., Xiong W., Rauh V., Wang S. (2012). Prenatal Polycyclic Aromatic Hydrocarbon (PAH) Exposure and child Behavior at Age 6-7 Years. Environmental Health Perspective, 120, 921–926.

BISPHENOL A EXPOSURE AND HEALTH RISKS

Edited by **Pinar Erkekoglu**
and **Belma Kocer-Gumusel**